BRICS Countries: Sustainable Water Resource Management and Pollution Control

Yiping Li · Hirok Chaudhuri ·
Otto Corrêa Rotunno Filho · Natalia Guseva ·
Faizal Bux
Editors

BRICS Countries: Sustainable Water Resource Management and Pollution Control

Challenges and Opportunities

Springer

Editors
Yiping Li
College of Environment
Hohai University
Nanjing, China

Otto Corrêa Rotunno Filho
Civil Engineering Program/Alberto Luiz Coimbra Institute for Graduate Studies and Research in Engineering—COPPE
Federal University of Rio de Janeiro (UFRJ)
Rio de Janeiro, Brazil

Faizal Bux
Institute for Water and Wastewater Technology
Durban University of Technology
Durban, KwaZulu-Natal, South Africa

Hirok Chaudhuri
Department of Physics and Center for Research on Environment and Water
National Institute of Technology Durgapur
Durgapur, India

Natalia Guseva
Department of Geology
National Research Tomsk Polytechnic University
Tomsk, Russia

ISBN 978-981-99-9580-6 ISBN 978-981-99-9581-3 (eBook)
https://doi.org/10.1007/978-981-99-9581-3

This Springer imprint is published by the registered company Springer Nature Singapore Pte Ltd.
The registered company address is: 152 Beach Road, #21-01/04 Gateway East, Singapore 189721, Singapore

Paper in this product is recyclable.

Contents

Brazil: Environmentally Integrated Basin Experiments (EIBEX) Driven by Hydrological Change, Sustainable Practices, and Water Security in Brazil

Otto Corrêa Rotunno Filho, Nilo de Oliveira Nascimento, Lígia Maria Nascimento de Araujo, Daniel Andrés Rodriguez, Afonso Augusto Magalhães de Araujo, Nelson Ferreira Fernandes, Alexandre Lima de Figueiredo Teixeira, Daniel Medeiros Moreira, Vanessa Lucena Cançado, Nívia Carla Rodrigues, Felipe Laender, Julian Cardoso Eleutério, Talita Silva, and Brigitte Vinçon-Leite

1 Introduction to Water Resources Management, Pollution Treatment, and Promotion of Ecosystem Services Provision in Brazil

Water resources management essentially focuses on promoting and enhancing the benefits that can be obtained from water resources through policies, which integrate structural and non-structural measures, taking into account the existence of limits to the services that can be expected from natural systems in a sustainable way. The

O. C. R. Filho (✉) · D. A. Rodriguez
Laboratório de Recursos Hídricos e Meio Ambiente (LABH2O), Programa de Engenharia Civil, Instituto Alberto Luiz Coimbra de Pós-Graduação e Pesquisa em Engenharia – COPPE, Universidade Federal do Rio de Janeiro (UFRJ), Campus Ilha do Fundão – Cidade Universitária, Caixa Postal 68540, Rio de Janeiro, RJ CEP 21945-970, Brazil
e-mail: otto@coc.ufrj.br

D. A. Rodriguez
e-mail: daniel.andres@coc.ufrj.br

N. de Oliveira Nascimento (✉) · V. L. Cançado · N. C. Rodrigues · F. Laender · J. C. Eleutério · T. Silva
Departamento de Engenharia Hidráulica e Recursos Hídricos, Programa de Pós-Graduação em Saneamento, Meio Ambiente e Recursos Hídricos, Universidade Federal de Minas Gerais (UFMG), Belo Horizonte, MG, Brazil
e-mail: niloon@ehr.ufmg.br

J. C. Eleutério
e-mail: julian.eleuterio@ehr.ufmg.br

Y. Li et al. (eds.), *BRICS Countries: Sustainable Water Resource Management and Pollution Control*, https://doi.org/10.1007/978-981-99-9581-3_1

existence of conflicts arising from the demands of different uses of resources and the exploration of new opportunities lead to the need for studies and research that consider the shared vision of the best-use policies. To meet those objectives, policies must consider short, medium, and long-term planning horizons, including strategies for mitigation and/or adaptation and resilience to extreme hydrometeorological events, inter and intra-annual climate variability, and long-term climate and hydrological change. The present chapter fits well in this framework, since the major focus is on some of the bases and principles that support scientific development and evaluation with respect to water resources management and pollution treatment in Brazil.

Population growth and global socioeconomic changes have increased demands for energy, water, food, and space. This expansion of human activities has intensified pressures on natural ecosystems that perform various eco-hydrological functions, such as water quantity regulation, erosion, and sediment input control, influencing the physical and chemical parameters of the watercourses (Blöschl et al. 2019). Climate change may be impacting hydrological processes at global and regional scales, influencing the water cycle, particularly rainfall regime, evapotranspiration, and air temperature, with impacts on vegetation cover, land and water uses, and eventually all the hydrological processes.

T. Silva
e-mail: talita.silva@ehr.ufmg.br

A. A. M. de Araujo
Departamento de Recursos Hídricos e Meio Ambiente, Escola Politécnica de Engenharia, Universidade Federal do Rio de Janeiro (UFRJ), Campus Ilha do Fundão – Cidade Universitária, Caixa Postal 68540, Rio de Janeiro, RJ CEP 21945-970, Brazil
e-mail: afonsoaraujo@poli.ufrj.br

N. F. Fernandes
Laboratório de Monitoramento e Modelagem Pedogeomorfológica (LAMPEGE), Departamento de Geografia, Instituto de Geociências, Universidade Federal do Rio de Janeiro (UFRJ), Av. Athos da Silveira Ramos, 274 - Cidade Universitária - Ilha Do Fundão, Cidade Universitária, Rio de Janeiro, RJ CEP 21941-916, Brazil
e-mail: nelsonff@acd.ufrj.br

L. M. N. de Araujo · A. L. de Figueiredo Teixeira
Agência Nacional de Águas e Saneamento Básico - ANA, Setor Policial Sul – SPS, Área 5, Quadra 3, Blocos "B", "L", "M" E "T", CEP 70.610-200, Brasília, DF, Brazil
e-mail: ligia.araujo@ana.gov.br

A. L. de Figueiredo Teixeira
e-mail: alexlima@ana.gov.br

D. M. Moreira
Divisão de Hidrologia Aplicada, Serviço Geológico do Brasil, CPRM, Av. PasteurCEP 22.290-240, Rio de Janeiro - RJ, Brazil
e-mail: daniel.moreira@cprm.gov.br

B. Vinçon-Leite
Laboratoire Eau, Environnement Et Systèmes Urbains (LEESU), École Des Ponts ParisTech, Paris, France
e-mail: bvl@leesu.enpc.fr

Understanding the interaction of natural variability and anthropogenic effects that governs hydrological change is critical to improving knowledge of the processes that govern the water cycle in connection with the rapidly changing human system. Integrating this knowledge into the planning and management of water resources appears to be one of the major challenges for water security and sustainable development. Regional and local characteristics of water availability and demand, ecosystem and water system resilience, governance policies and instruments, among others, condition water security.

In general, the concepts of water security include disaster risk assessment, mitigation, and guarantee of supply and access to water, focusing on the most vulnerable people. But they also incorporate the economy, material goods, and ecosystems. Ensuring water availability for human supply and for productive activities is a matter of water security that can be affected by floods and droughts, or by inadequate exploitation of the resource, degradation of its quality, and loss of capacity of the resources in ecosystems for their regulation (Mason and Calow 2012). According to Vörösmarty et al. (2010), nearly 80% of the world's population is exposed to high levels of threat to water security.

Although there have been escalating discussions whether water is a public good or a market commodity, it is recognized that water is initially a fundamental input to any socioeconomic activity, such as agriculture, energy, transport, industry, and even production of healthy human resources, playing the supportive role toward economic growth. Spatiotemporal management of water quantity and water quality requires investments in water transfer, water diversion, storage, treatment, and land use planning, among other anthropic interventions (Siska and Takara 2015). It is recognized that the vulnerability of water resources systems to floods and droughts will increase, and the trade-offs among reservoir releases to maintain flood control storage, drought resilience, ecological flow, human water demands, and energy production should be urgently reconsidered (Ehsani et al. 2017).

These issues must be assessed in long-term planning of water resources. The potential impacts of alternative policies are explored through the use of numerical models to simulate different scenarios. Such approach relies on the capacity of model structure to simulate patterns and functions of the basin system and on feasibility of assessing model uncertainties. For this, the availability of long-term information that allows understanding the natural system is essential. Experimental basin studies conducted in representative areas are one of the most useful tools to assess this issue.

In order to explore water resources management in Brazil a few case studies are selected. The main scientific bases for addressing such case studies are: (i) the strong integration between intensive experiments, continuous monitoring, and mathematical computational and analytical modeling for solving environmental problems; (ii) interdisciplinarity and inter-institutional collaboration; (iii) strong integration with undergraduate, post-graduate strictu sensu, and graduate latu sensu; (iv) scientific-technological extension projects.

Furthermore, a set of scientific and technological macro-issues was identified whose solutions can contribute significantly to BRICS science and its social impact. The macro-issues are:

(i) Demand for scientific and technological training by the BRICS technical staff in private companies, state-owned companies, and public agencies, which seeks to reduce the waste of resources, purchase or proposition of expensive technologies and solutions, and general inefficiency of the economy.

(ii) Environmental vulnerability: climate change, population growth, and environmental pollution all result in increased pressure on natural resources and increased vulnerability of population (especially the corresponding poorest portion) to climatic extremes and ecological accidents or disruptions. The importance of environmental monitoring, quantitative diagnostic models, and models capable of accurately predicting the effects of changes in climate or ecosystems or of environmental accidents is clear. A major importance has been identified for short-term forecasting models of severe events in the atmosphere, in river basins, and at sea to improve environmental risk management with a particular focus on poor populations living in high-risk areas, as well as simulation and long-term forecasting models that are essential tools to guide land use and correct management of terrestrial ecosystems.

(iii) Nonlinearity and feedback mechanisms in nature: in its present stage, science is still not able to fully understand the operation of complex nonlinear systems (i.e. with a large number of dimensions), as is the case of flows in general, including various types of undulating flows and turbulent flows, the physical climate system, chemical chain reactions that determine the concentrations of pollutants in water, soil and air, and ecosystem dynamics. Basic studies in applied mathematics, dynamic systems, and fluid mechanics, among others, are fundamental for improving the scientific capacity to model and predict. In particular, it is intended to follow and deepen the tendency to replace empiricism (in the sense of ad hoc solutions and valid only for very specific situations or regions) by models based on solid scientific concepts.

(iv) New measurement and monitoring methods: inevitably, item (iii) above represents an ambitious and long-term program. Often, in order to deal with the complexity of environmental problems and/or deepen the knowledge of nature, it is necessary to resort to field measurement and monitoring. The question of environmental data cannot be relegated to the collection of sparse information (in time and space), and often erroneously measured or highly uncertain. The measurement of relevant quantities in a systematic, continuous, and accurate way requires new technologies and methodologies, development of robust and reliable systems, quality control techniques, and is an integral part of the proposed methodological approach to environmental problems.

(v) Low capacity for implementation of water resources management instruments in the states and municipalities, due to the reduced number of employees and high turnover of effective technical staff, is to evolve to technical and managerial improvements through adequate training in face of the complexity of the environmental problems in order to contribute to the practical application of the acquired knowledge frequently supported by public funds.

Additionally, in addressing the integrated management of water resources, it makes sense to adopt the river basin scale as the adequate territorial unity. This perspective is in line with the recommendations by the Brazilian Law 9433/97 or so-called the Brazilian Water Law (January 8, 1997) and with the creation of the Brazilian National Water Agency (ANA) in 2000.

To effectively address the multidisciplinary approach that the management of water resources and environment demand, a variety of scientific and managerial new developments are required, involving, among others, the following topics:

- the acquisition and interpretation of data and information on the physical processes and geoenvironmental characteristics related to the water cycle at scales compatible with the unit of analysis;
- understanding the relationship between the natural space–time scales of observation and modeling of hydrometeorological phenomena;
- understanding the functioning and behavior of basins of different geomorphological characteristics as well as under different forms of anthropization;
- comprehension of biogeochemical flows associated with the water cycle;
- identifying changes on the hydrological cycle and on water quality due to catchment land use characteristics;
- conceiving river basin management programs aiming at recovering hydrological processes and increasing resilience to climate change;
- estimating river basin recovery costs and investigating funding alternatives,

Under this very broad vision, the following scientific issues to receive priority attention are here-in identified, namely: (i) global climate change and its regional effects, (ii) hydrology and meteorology at the mesoscale, (iii) surface-atmosphere interactions, micrometeorology and hydrological river-basin processes, including the genesis of catchment flows through the analysis of surface-vadose zone-groundwater interactions, (iv) quality of water, air and soil, and (v) management of water resources and pollution treatment.

This chapter establishes an academic framework for the evaluation of the hydrological, hydrometeorological, hydroclimatic, and environmental behavior in catchment areas in the highlands of the states of Rio de Janeiro and Minas Gerais. The framework illustrates research works developed in Brazil, covering urban, agricultural, and forested regions, based on the consolidation of the knowledge of hydrological studies developed in Brazil over various national river basins, including the Amazon basin. The focus is on water quantity and water quality accordingly to the main topic addressed jointly by the group—*BRICS Water Resources and Pollution Treatment*.

The use of remote sensing and GIS in conjunction with computational numerical models is emphasized, with a view to the development and evaluation of methodologies and hydrological and atmospheric parameterizations in the river basin scale.

Additionally, it is argued to be essential to articulate obtaining basic field data from in situ monitoring and remote sensing technologies, enabling multidisciplinary studies on modeling of water resources, with emphasis on the physical and water quality aspects of the understanding of the hydrological cycle and evaluation of climate change. This perspective is integrated with a social vision of the physical system of the basin, providing updated information to society focusing on water security.

This chapter follows the guidelines as highlighted below:

- implementation of hydrometeorological models that address surface and groundwater hydrological phenomena, aimed at forecasting flows and water levels during critical events, floods and droughts, as well as allowing the methodological development of water quality diagnostics via modeling rainfall-runoff in watersheds and also performing water balance tests in basins and simulation of hydrometeorological events incorporating four environmental components: soil-air–water-vegetation;
- conceiving scenarios of water and soil conservation measures at urban and rural environments, assessing the associated benefits for water security and conceiving water management frameworks that can promote the implementation of those measures, including assessing costs and funding alternatives.

1.1 Selected Case Studies

As previously mentioned, selected case studies for Rio de Janeiro and Minas Gerais states are examined:

(i) the Piabanha river watershed in the mountainous region of the Rio de Janeiro state, a sub-basin of the Paraiba do Sul river basin (Fig. 1);
(ii) the Serra Azul catchment, located in the transition Atlantic forest-savanna biomes of central Brazil, in Minas Gerais State (Fig. 2).

In the first case, the emphasis is on improving the characterization of the hydrometeorological regime related to the dynamics of floods of the Piabanha river. The experiments take place at experimental basins with heterogeneous land use and vegetation cover all nested in a representative basin.

The second case study focuses on the use of economic instruments for integrated natural resource management tools. The focus is on catchments located in metropolitan areas, equipped with reservoirs, and that are strategic sources for drinking water supply. Constructing and simulating future land-use scenarios using a business model as usual approach or soil and water conservation strategies is one way to clarify threats to water security in those metropolitan areas. Other approaches include the assessment of the potential economic, social, and environmental benefits in terms of protecting water sources, with additional benefits such as flood control, the promotion of biodiversity in urban and peri-urban areas and the creation of opportunities for social exchanges.

The implementation of an environmental program is discussed through the projection of land use scenarios, based on main regional land use trends, and the introduction of protective practices and environmental recovery approaches. Costs of setting up

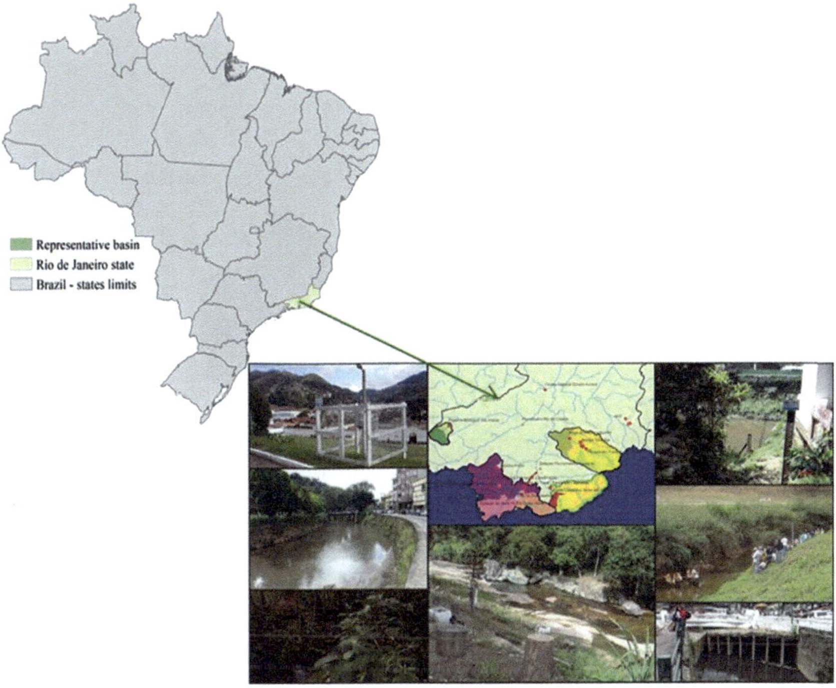

Fig. 1 Location of the Piabanha river basin and the representative area—state of Rio de Janeiro—southeastern Brazil (upper left panel); site-specific land use and land cover—forested, agricultural and urban areas (lower right panel)

and handling the measures, in addition to the possible sources of financing, considering their relevance in the viability and acceptance of the program, are addressed. The case study area is the Serra Azul catchment, one of the main water sources for the Metropolitan Region of Belo Horizonte (located in Southeast Brazil), which is equipped with a reservoir for drinking water supply.

Basins in the southeastern Brazil present a hydro-meteorological and hydrological connection with the Amazon basin (6,112,000 km^2). The Brazilian part of the Amazon basin occupies about 61% of the terrestrial surface of Brazil, noting that approximately 5,217,423 km^2 or 85% of the total area of the basin is located in Brazilian territory (8,516,000 km^2). Connection is evident through circulation in the atmospheric system, whereby the Amazon region plays a role of atmospheric moisture source, and through the economic system. In addition, it should be taken into account climate change scenarios and the behavior of floods, droughts, and social impacts that are produced due to the operation of the interconnected hydroelectric power generation and distribution system in Brazil. In this sense, the understanding of the mechanisms and pattern behaviors of wetlands, for example, could be addressed as part of the overall water and energy balance for Amazonia and for the southeast of Brazil.

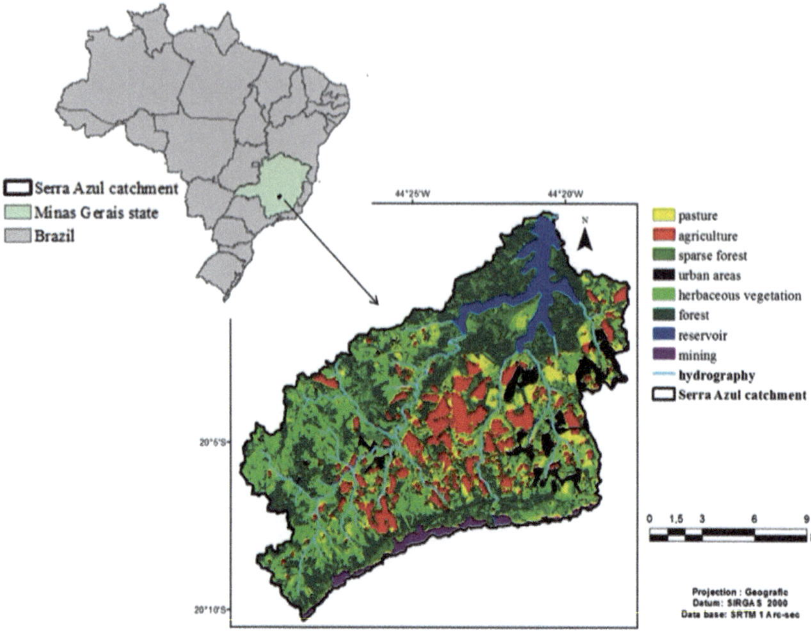

Fig. 2 Serra Azul watershed—Minas Gerais—southeast Brazil—forested, agricultural and urban areas

Indeed, as revealed by the connection between the Amazon basin region and the mountainous area of Rio de Janeiro, a broad and an integrated overall national water resources system that stimulates the integration of the diversity of entities quite active in the area of water resources and pollution treatment across the country is required. Furthermore, it is worthwhile to point out the importance of the development of satellite platforms, sensors, and technical tools for acquisition and analysis of a variety of in situ and notably spatiotemporal remotely sensed datasets, namely exploiting their potential for monitoring water and climate change, in order to solve general and specific problems nationwide.

Such broad hydrometeorological and hydroclimatic connections in Brazil are depicted in-here by the investigation of the relationship that exists between the selected case study referred as Piabanha river watershed and the worldwide well-known Amazon river basin. It is expected that the information and related thematic issues previously reported and still to be explored in this text should highlight the challenges to build a national water system to support the decisions to be made regarding water resources management and pollution control, focusing on environmental protection and sustainable social development.

It should be noted that the approach envisioned along the scope of this text is to explore the local and small scale to uncover connections and challenges in upper and larger scales. Therefore, the methodological framework is conceived to allow procedures to be developed and evaluated at the scale of representative basins, which, in a very broad sense, could be thought to be like a kind of experiment to be conducted at the laboratory scale. In this perspective, computer model techniques are addressed and tested and particular areas are monitored for retrieval of remotely sensed and in situ measurements, which are then jointly employed for evaluation to develop and achieve recommendations and solutions to support watershed management policies and practices.

1.2 General Background About Water Resources Management in Brazil

Brazil encompasses an area of approximately 8,516,000 km^2 and has about 12% of the world's surface water availability (ANA 2009).[1] Although abundant, 80% is found in the Amazon Region, with low population density, and, consequently, less water demands. Part of those outflows is allocated to a wide spectrum of water uses, even in drier years. Surface water availability in Brazil is estimated to be around 78,600 m^3/s or 30% of the mean flow, with the Amazon Basin contributing 65,617 m^3/s (ANA 2017).

The water resources institutional framework in Brazil, as proposed by the Brazilian Water Law of 1997 (Law no. 9433/97), has the goal to ensure water security regarding quantitative and qualitative aspects of the human needs, animal watering, and economic sectors, considering multiple uses. The current water policy in Brazil follows the National Water Resources Policy (PNRH) through the National Water Resources Management System (SINGREH). The National Water Agency (ANA), which was created by the Federal Law no. 9984 of 2000, is the federal agency responsible for implementing this policy. In addition, at the state level, the water resources management is conducted by 27 state government framework gathering state water resources councils, environmental protection agencies, and river basin authorities.

Water resources availability and demands spatial distribution

The spatial and temporal river water flow variation in Brazil combined with some drivers, such as the intense water withdrawal and raw or poorly treated sewage released into rivers, cause water balance instability in many regions. According to the National Water Supply and Sanitation Information System—(in Portuguese, Sistema Nacional de Informação Sanitária—SNIS) (SNIS 2018), 46% of the sewage are treated and only 53% of the Brazilian people have access to a wastewater collection system. Figures 3 and 4, which were extracted from the Brazilian Water Report (ANA 2017), depict a spatial overview of the quantitative and qualitative water balance representation in Brazil. The first map (Fig. 3) shows the vulnerable regions in the

---

[1] Considering the flows from foreign territory entering the country, this total water availability reaches values of around 18% of the world's surface water availability (ANA 2009).

territory due to water availability versus demand problems. The second diagram (Fig. 4) represents the water qualitative stress for each river stretch.

For much of the last century, interventions in the water sector in Brazil favored solutions focused on expanding water availability, quantitatively speaking, such as the construction of water storage reservoirs. As an example, according to ANA (2017), Brazil had 19,361 artificial reservoirs mapped in 2016, and the evolution of

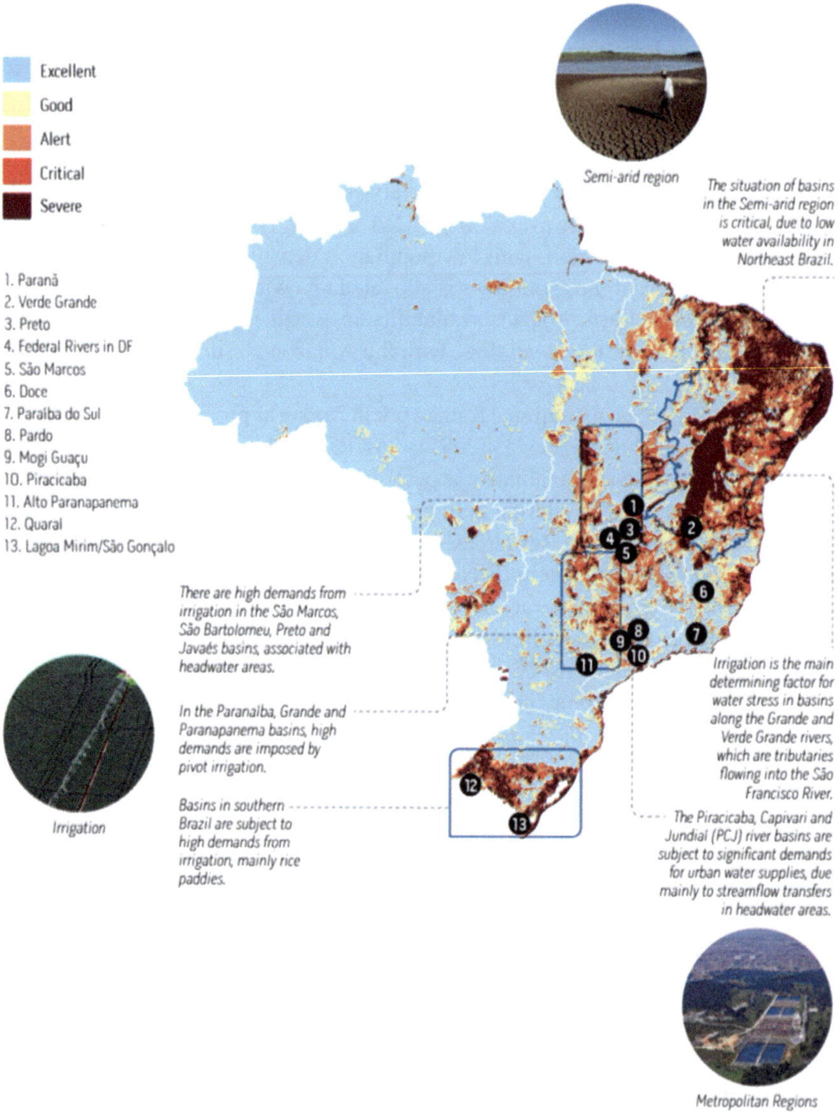

Fig. 3 Quantitative water balance in Brazil (ANA 2017)

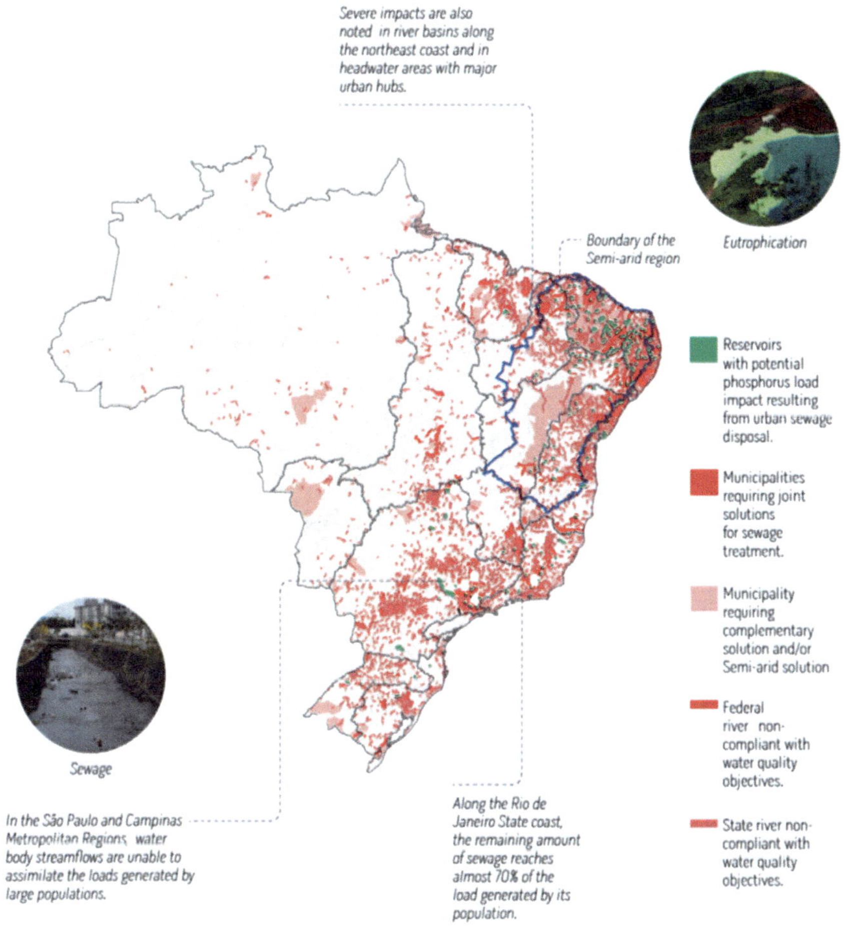

Fig. 4 Qualitative water balance in Brazil (ANA 2017)

the water storage capacity in the country was intensified after 1950, especially with the addition of large volume of reservoirs constructed by the Brazilian hydropower sector (Fig. 5).

However, due to the great increase in water demand in the country (Fig. 6) and the intensification of conflicts, the importance of acting not only in supply but also in the rational management of water resources demand was verified.

A resource is efficiently allocated when it is used to maximize its value. In the case of water, a large number of different and partly competing uses can be envisaged, for example in industry, agriculture, and tourism. Achieving an efficient allocation of water resources can mean changing water use within or between sectors, such as from low to high value crops in agriculture, or use in agriculture as opposed to industrial use.

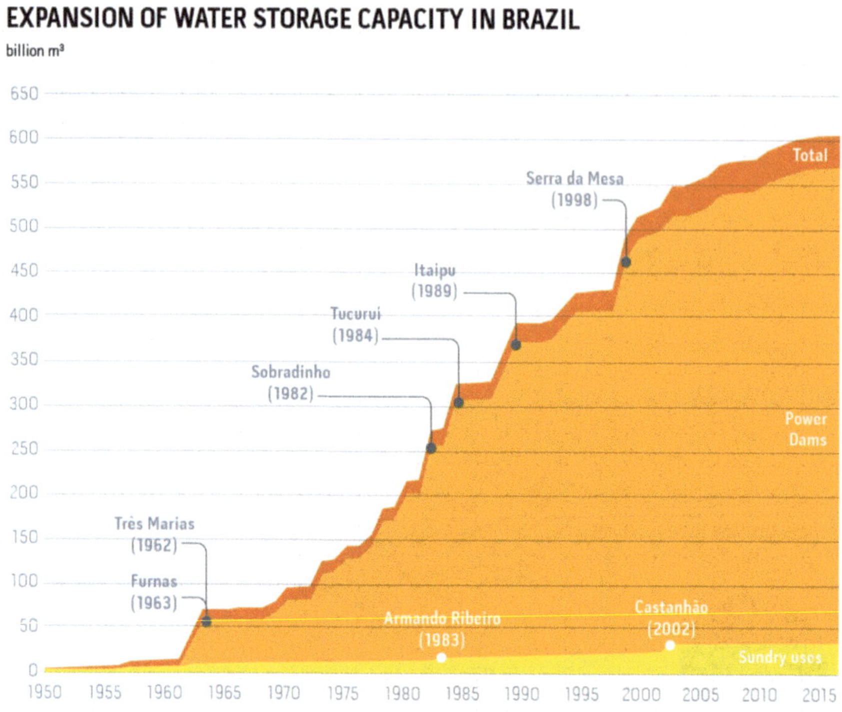

Fig. 5 Evolution of the capacity of water reservation in Brazil (ANA 2017)

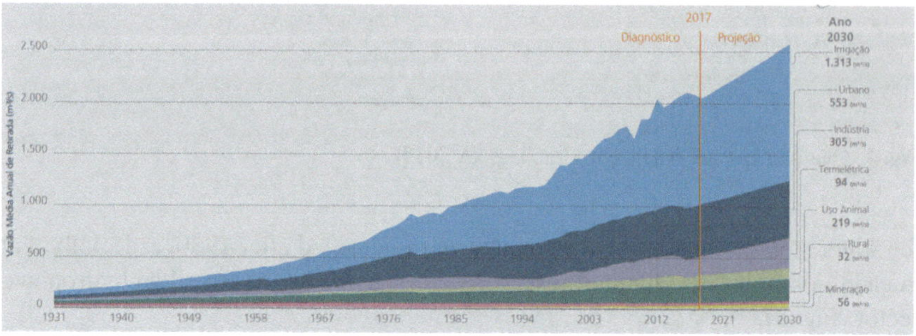

Fig. 6 Water demand in Brazil over the years (ANA 2019)

According to the Brazilian Water Report (ANA 2017), the Water Resources National Policy has provided advances related to each legal water tool or mechanism. The 12 Interstate Water Resources Plans up through to 2019, for instance, cover an area corresponding to 54% of Brazil. In Brazil, the total number of water grants for all sectors accounts for 13,657 in federal rivers and 101,435 in state rivers. In addition, the water quality targets (classes), which may vary along the different river stretches, were adopted in some basins. Water charges, which are supposed to encourage rational water use, summed up, until 2016, USD 70.7 million and USD 250.4 million in Federal and State jurisdiction rivers, respectively. Finally, the Water Resources Information System—SNIRH (www.snirh.gov.br), which is a platform created and maintained by the National Water Agency (ANA), in a partnership with more than 50 Brazilian institutions, provides a comprehensive view of the water resources situation in Brazil.

Despite all this effort and progress with the implementation of the Water Resources Law, it is noted that the formal instruments of water resources management in Brazil have sometimes not been enough to adequately manage conflicts or, in some cases, to allow the promotion of actions to anticipate potential future problems, such as changes in climate. Data from the Brazilian Water Resources Report—2017 (ANA 2017) reveal the large number of people affected by recent water crises, where several systems, notably in the most critical areas of the country, collapsed due to climatic factors, evidencing low water security. According to ANA (2017), from 2013 to 2016, 48 million people over a population of 200 million inhabitants were affected by floods and droughts in Brazil. A total of 4,824 drought events associated with human damage were quantified, almost three (3) times more than floods (1,738). In this context, a wider concept of water security proposed by the United Nations[2] has been considered in the implementation of the Brazilian Water Resources Policy.

Water Security

According to Srinivasan et al. (2017), water security is a multi-faceted problem, going beyond mere balancing of supply and demand. Early attempts to quantify water security relied on static indices based on approaches that have failed to acknowledge that human action is intrinsic to the water cycle. Human adaptation to environmental change and increasing spatial specialization in the modern world require a more flexible and dynamic view of water security. Starting from more fundamental principles, and through application of simple water balance concepts to human-impacted water systems, the cited research work first developed a set of indicators for water insecurity. The authors subsequently offered an approach to model these indicators as outcomes of coupled human-water systems to anticipate watershed trajectories under anthropic impact in order to predict water insecurity and inform appropriate action. Thus, far from being a static index, water security means a "safe operating subspace" within a three-dimensional space that maps physical resource availability, infrastructure, and economic choices.

The Brazilian Water Security Plan (ANA 2019) defined the Water Security Index (WSI) (in Portuguese, ISH, as shown in Fig. 7) for the whole country, covering the main dimensions of water security, namely human, resilience, ecosystem and economic. The spatial distribution of WSI in the territory revealed the occurrence

[2] "The capacity of a population to safeguard sustainable access to adequate quantities of acceptable quality water for sustaining livelihoods, human well-being, and socioeconomic development, for ensuring protection against water-borne pollution and water-related disasters, and for preserving ecosystems in a climate of peace and political stability."

of areas where the level of water security is extremely low. It can be seen that most of the northeast region shows low or minimum WSI, due to low precipitation rates and consequently low water availability. Minimum and low WSI occurs also in some specific areas in the southeastern region, such as the big cities of Rio de Janeiro and São Paulo and their surroundings. In the extreme south of the country, the low WSI is the result of intense water use for agriculture irrigation, in special irrigated rice.

The adoption of alternative mechanisms for the formal management of water resources is of fundamental importance for anticipation and action in these critical situations (ANA 2017). The allocation of water resources is related to a more efficient and sustainable distribution of water, especially during crisis situations. In response to these challenges, more complex and modern forms of water allocation in river basins are required.

Instead of a simple set of fixed rules, modern allocation plans should include or rely on scenarios that project how water use can respond to climate change, oscillations

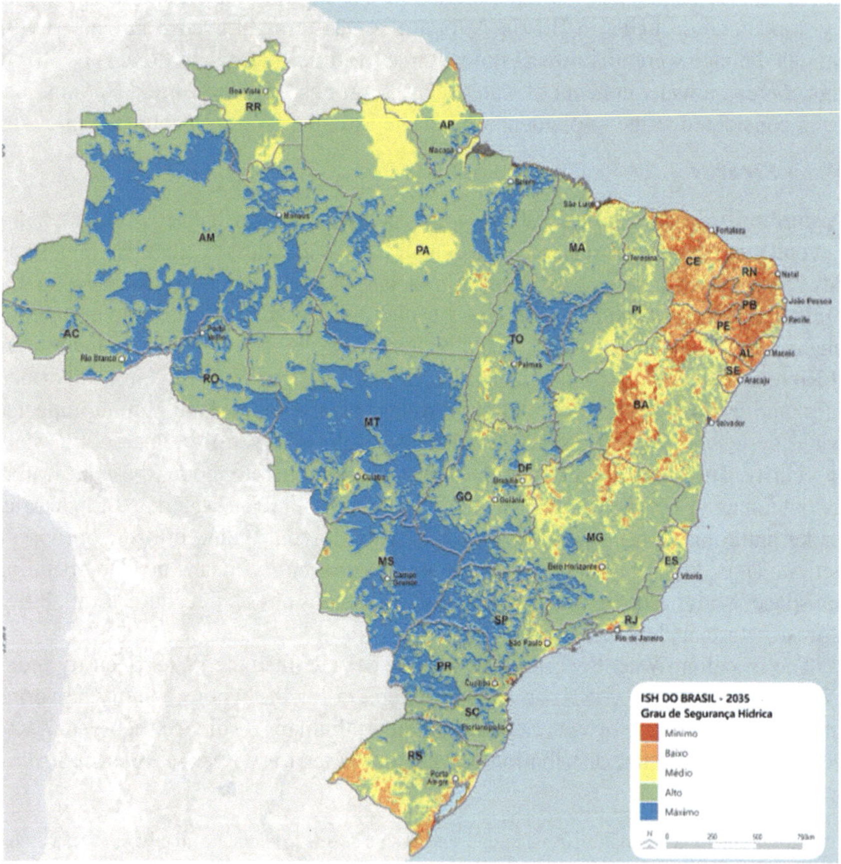

Fig. 7 Water security index (WSI) in Brazil (ANA 2019)

in economies, water price incentives, and options for sharing benefits of water use. These approaches should increasingly consider the unpredictability of future water resource scenarios in river basins, as pointed out in the studies by Speed et al. (2013).

In order to address shortages and ensure that water is available for all uses, negotiated allocation agreements may be drawn up for specific reservoirs and other systems. Water allocation is a management strategy deployed to regulate uses in water systems assailed by severe droughts, in emergency situations or with a strong likelihood of disputes. Between 2014 and 2016, 34 water allocation agreements were dealt by ANA in Brazil with state management entities, dam operators, and water resources users, working closely with basin committees (ANA 2017).

Integrated environmental management of water resources: ecosystem services provision

Increasingly common are approaches that conceptually and analytically emphasize ecosystem services as a form of protecting the environment. Clarifying what these services are and their importance for human well-being as well as the very survival of living beings are essential information for environmental education, and new tools seek to incorporate ecosystem services within the scope of socio-environmental policies and programs. These new approaches question the existing institutional restrictions, since, in rare exceptions, are unpaid services on the market with a fragile legal base that guarantees their appreciation.

Successful experiences of encouraging the provision of environmental services in watersheds that are sources of water supply are viewed with great enthusiasm by researchers, environmental analysts, and public managers, both for the potential of environmental and financial gains, since investing in watershed management can be less expensive than developing new water sources or treatment structures (Neto 2008; OECD 2010). Four main arguments are used in the discussion on solutions for the provision of ecosystem services and their necessity or justification: legal, economic, political, and ecological. These are subjects that interact with each other in the construction of an effective arrangement to stimulate the offer of services. A brief analysis of these dimensions is convenient for understanding the problem that involves ecosystem services.

The creation of strategies to encourage the provision of ecosystem services would be justified from the perspective of the prescriptive lower and upper power limits of the state, both defined by the degree of absence of legislation that effectively guarantees the provision of ecosystem services in quantity and the quality necessary for our well-being with regard to its reach of power in resolving all existing environmental conflicts. The multiplicity of different situations, transactional costs involved, and difficulties of inspection would limit public power for the protection and resolution of environmental conflicts. It should be noted that here the criticism would be related mainly to a type of state mandate, predominantly protector-repressive, contradicting a mandate that would have a promotional function. The first type of mandates would seek to discourage undesirable actions for society, by making it impossible to implement them, by making it difficult to execute them or by making them disadvantageous or inconvenient. By contrast, mandates with a promotional function would seek to provoke the exercise of compliant acts, seeking to make the socially desired action more necessary, easy, and advantageous (Bobbio 2011; Tejeiro 2013).

Mechanisms to stimulate the provision of ecosystem services include the use of economic instruments (EI) in environmental management and planning. These procedures consist of tools such as the imposition of taxes for polluting activities, compensations for the provision of ecosystem services (monetary or not), and the possibility of negotiating permits and authorizations to negatively impact the environment.

The concept of "economic" is related to a way of analyzing the factors that interfere in the behavior of the agents or social actors and would not be restricted to "financial," "market" or "monetary." Even the term "market" for ecosystem services encompasses different contractual arrangements according to each country (Salles et al. 2006). Because most environmental services are positive externalities (they generate benefits not recovered by those who produce them), they are not effectively incorporated into the decision-making process as a relevant and determinant input variable.

It is assumed that the use of the EIs would make it possible to reach the objectives defined by the public environmental policy via modification of the behavior of economic actors. With the use of appropriate tools, agents can react by modifying behaviors that would negatively impact the environment; because it is sought to influence the psyche of those governed by the mandates, inducing a certain type of behavior (Bobbio 2011; Forest Trends 2015). The Conference of the Parties on Biological Diversity of the UN (COP 10/CBD), held in 2010, was a milestone in valuing economic arguments for biodiversity conservation and the need to communicate interdependencies among human well-being, economic development, and natural capital for the definition of effective conservation strategies.

A relevant economics contribution would be in how to incorporate ecosystem services into the "economic environment," and what would be the possible and appropriate arrangement in order to reconcile individual and collective decisions on the use and occupation of the soil on a sustainable basis. An important question that arises in terms of how to enable the transfer of resources between actors—creation of markets, subsidies, taxes, financial remuneration for the provision of services, in-kind compensations (goods and services delivered without the involvement of money, as the possibility of access to training and new technologies). Secondly, on how to value ecosystem services, considering that the economic system is only interested on what has "value," and the way in which individuals and society perceive this value is decisive for achieving the desired results in an environmental program.

Defining the value associated with environmental changes has obvious difficulties because of the great connection between ecosystem services and the systems that offer them (Seppelt et al. 2011). Many authors emphasize the importance of undertaking an integrated and comprehensive assessment, for which a multidisciplinary approach is required. However, there are differences in the value classification system, which generally encompass economic, biophysical or ecological and sociocultural values (Groot et al. 2002; Groot 2006; EPA 2009; Castro et al. 2014; Kenter et al. 2015). The methodologies for capturing these classes of values are diverse. In the case of economic values, in the absence of markets for ecosystem services, techniques seek to capture preferences by constructing hypothetical markets for environmental goods or by observing consumer behavior in other markets that would be related to the environmental good or service (Groot et al. 2002; Millennium Ecosystem Assessment 2005; Kenter et al. 2015). An example is the hedonic pricing method, which uses statistical techniques to break down the price paid for a service into the implicit prices for each of its attributes, including environmental attributes such as access to recreation or clean air.

On the other hand, the need to evaluate ecosystem services, and at what level of detail, is closely linked to the objective to be achieved and the tools that will be used to generate or protect the environmental benefits. Authors argue that valuation techniques would be relevant to help create an idea of the value of the service provided, even if this value is far from the current negotiations; others believe that it will suffice for both sides to be satisfied, so that the market can advance (for a comprehensive literature review on these issues, please see Neto 2008). May (2003) indicated that the assessment of environmental services was proposed as the first step toward the definition of incentives for its maintenance. Although their researches also suggest that the opposite may be even more valid (i.e., once society undertakes the creation of an incentive for the conservation of nature via political processes, it ends up creating value in a context where the market had not previously actuated). This was observed in many cases of payment for ecosystem services (Neto 2008). Wunder (2005) argues that having a good idea of the opportunity costs of potential service providers may be much more useful, and much simpler, for setting up ecosystem services in comparison to developing studies, sometimes complex, to assess the environmental services in question. Salles et al. (2006) argued that complete monetization of environmental values is not necessary in practice since it is often enough to compare the cost of payments for ecosystem services with that of a technological solution. Numerous factors determine price formation in payment for ecosystem services (PES) business models, such as the degree of competition, both in supply (price at which the seller is willing to sell) and in demand (price that buyers are willing to pay), and, in this case, the "current price" is hardly coincidental or superior to the "environmental value" (Bracer et al. 2007; Neto 2008). This is another issue concerning the markets for ecosystem services.

In the absence of this valuation in programs of payment for ecosystem services, general values are often defined for environmental compensation or service provider remuneration (which also has the advantage of having a lower transactional cost), such as, for example, using, as a reference, an estimate of the opportunity cost (an average value of the revenue that the owner would have ceased to obtain by dedicating part or all of his property to the provision of ecosystem services to the community), or the average cost of the recovery measures that would be required in the absence of the environmental service. To this matter, Salles et al. (2006) pointed out that the success of the Bush Tender (Salzman 2005), the payment program for ecosystem services in Australia, is due more to the method of calculating the value of biodiversity, performed for each property that applied to benefit from the program, than the form of payment for auctions.[3]

Salles et al. (2006) listed four types of payments, which have different transactional costs. The simplest would be a fixed general and uniform subsidy. It is the least expensive to organize, but on the other hand, does not distinguish the beneficiaries who could provide the most interesting ecosystem services at that price. There may then be direct negotiation with service providers (usually landowners) to find a compromise tailored to each case. This approach is expensive and does not allow for competition between owners, and at the same time does facilitate a coordinated and integrated approach. This problem can be solved, in turn, through the organization of reverse auctions,[4] as in the case of the Bush Tender, where competitive dynamics develop among farmers. Salzman (2005) believes that this approach is particularly suitable for monopsony situations (only one buyer), unless the sellers are few and there is collusion among them. A final option would be to *pay* a third-party representative of service producers, i.e., a local authority, a non-governmental organization (NGO) or an institution set up for that purpose. This approach can help create a climate of trust, but it does not solve the issue of the amount to be paid for each specific action.

In Brazil, there are several initiatives for cash payments (direct monetary flow), which have different arrangements. These almost never rely on a valuation of ecosystem services *strictu* sensu, but, instead, on reference values associated with a scoring system which may be more or less simple and encompass one or more analysis criteria. This system can incorporate aspects such as an area with native vegetation, biotic quality of the preserved remnant, phytogeographic region, opportunity cost, management techniques, and conservation and environmental protection measures used, among others. It is mainly a remuneration for conservation and also for good practices, which potentially have environmental benefits (Guedes and Seehusen 2011; Hercowitz 2011; Santos et al. 2012; Forest Trends 2015).

After the initial exaltation of payment programs for ecosystem services, seen as an alternative to command and control techniques in environmental policy, criticisms of the model intensified and many studies began to less emphasize the question

---

[3] Bush Tender was a pilot program developed by the Natural Resources Department of the State of Victoria (Australia), between 2001 and 2003, to conserve remnants of native vegetation on private farms beyond the legal requirements and obligations. The program is based on a robust valuation methodology, using the reverse auction technique. Government officials visited the properties enrolled in the program to assess their diversity and their biological value. The owners then submit their proposals, including a native vegetation management plan and an annual award (bidding). According to the offers of the owners, the beneficiary properties were chosen, based on the best cost–benefit ratio considering the valuation of biodiversity previously assessed by the technicians. With the use of the reverse auction, it was possible to protect areas of great biological value at a value lower than what the government was willing to pay (Salles et al. 2006; Hercowitz 2011).

[4] It is called a reverse auction because it works opposite to a normal auction. In the reverse auction applied to ecosystem services, the object to be auctioned is defined (for example, the contracting of a number of hectares for the conservation of riparian forests) and sellers make their offers. Whoever offers the product (in this case, conservation areas) at the lowest price wins the auction and is hired under the ecosystem services payment program.

of financial compensation and to focus more on integrated solutions (Seppelt et al. 2011; Grima et al. 2016). These alternatives are based on the building of partnerships between ecosystem service providers, public power, entities and companies operating in the river basin, and other stakeholders, in a continuous process that guarantees the engagement of communities involved in the environmental program, awareness of the effect of their practices and alternative practices in the landscape and the environment, with the aim of medium and long-term self-regulation (see, for instance, Ostrom 2010).

Arguments against the use of economic instruments emphasize elements such as the risk of "commodification" of the environment, the subordination of man, and the nature of mechanisms of the market economy (Polanyi 2001 apud Wjuniski and Fernandez 2010; Superti and Auberti 2015), the domain of economist instrumental rationality in the decision process to the detriment of other criteria (Wjuniski and Fernandez 2010) and the incorporation of values as altruism, existence, and patrimony only in perspective of personal satisfaction or gain, present in the individual maximization approach (Kenter et al. 2015). Superti and Auberti (2015) reported that environmental conservation by market mechanisms is a myth in Brazil, indicating that the institution of environmental markets is occurring more at the level of scientific and technical discussions, and less in practical applications. According to the authors, use of payment tools for ecosystem services is mainly performed as an income transfer device.

Enthusiasts of the ecosystem services approach, which are mainly aligned with the current ecological economics, argue that when analyzing ecosystem functions, not only economic aspects but also environmental, natural, and cultural heritage are evaluated (Souza et al. 2020). Although the existence of markets is important, their non-regulation would be inadequate for the allocation of goods and services provided by nature (Souza et al. 2020; Irigaray 1993). Nusdeo (2013) pointed out that although payment for ecosystem services is an economic instrument, it is not necessarily a market instrument, because it is often based on strong governmental action or through recommendations from third-party entities. Schomers and Matzdorf (2013), reviewing 457 papers on payment for ecosystem services (PES) experiences, reported that PES is a multifaceted term and many diverse conservation approaches are published under this "label," noting that one-third of all publications focus on Latin America, and private "market negotiations" among social actors are little described in literature.

In any case, there is a relative consensus that it is not a question of privileging a specific technique—command and control, imposition of fines and taxes, persuasion of economic agents, payment—but analyzing those considered most appropriate for each situation. This means considering not only physical and biotic aspects, but also the social reality of the territory, understanding and considering the social processes and the actors who have set aside a space to be preserved or not.

Given the general overview provided about water resources in Brazil, the next sections present in-depth background discussion about hydrological assessment of water resources. Attention is dedicated to the applications herein presented through two case studies developed for relatively small basins located in Rio de Janeiro and

Minas Gerais states, namely Cases 1 and 2, respectively. Both basins belong to the southeast region of Brazil and therefore are influenced somehow by the hydrological and hydrometeorological behavior of the Amazon basin, notably with respect to the water and energy balances, as previously emphasized in the introductory section.

2 Case Study 1: Hydrological Patterns in Piabanha River Basin—Rio de Janeiro—Inserted in the Paraíba do Sul River Basin—Southeastern Brazil

The approach used for developing this case study recognizes the complexity of the physical processes that occur in nature (Dooge 1986, 1997). Therefore suggestions are made to simplify the representation of these processes in a hydrological model, trying to achieve a model behavior as close as possible to nature behavior. Rainfall-runoff models and remote sensing technology provide support to better understand some of the key issues for stimulating new reflections on hydrology and soil physics. Rainfall-runoff models are used to achieve a better understanding of the land phase of the hydrological cycle. Progress in this area has been through the development and use of two broad classes of models: lumped and distributed. However, many issues remain, such as how to include new input data sources as well as how to improve methods for calibration, validation, parameterization, and upscaling (Rotunno Filho 1989, 1995; Rotunno Filho et al. 1996; Araujo 2006; Libos 2008; Getirana 2009, Getirana et al. 2009, 2010; Xavier et al. 2010; Silva et al. 2010, 2012; Xavier 2012; Lucena et al. 2013; Medeiros 2016; Medeiros et al. 2016; Rotunno Filho et al. 2019).

Aitken (1973) studied 17 applications of rainfall-runoff models, published in the literature, highlighting that in many of them the structure of the models used resulted in flow overestimation or flow underestimation. More recently, Troy et al. (2015) proposed to review hydrological modeling and corresponding applications under the perspective of socio-hydrology. O'Donnel and Canedo (1980) presented a brief survey of the main drivers of uncertainties when modeling a watershed. These uncertainties could be grouped as.

(1) Basin data records:
(a) errors in collection of raw data;
(b) reduction of point data to spatial averages;
(c) reduction of continuous data to temporal averages;
(d) estimation procedures for non-directly observed variables (e.g. evaporation);
(2) Basin model structure:
(a) imperfect knowledge of basin processes;
(b) approximations to achieve tractable model representations of basin processes;
(c) model lumping of spatially variable basin properties (e.g. infiltration capacity);
(d) sequential one-at-a-time model treatment of processes which occur concurrently in a basin;

(e) omission of basin processes considered unimportant to a given modeling purpose;
(3) Model calibration:
(a) choice of objective function;
(b) length of calibration record's sample;
(c) calibration variability over different samples;
(d) criteria for declaring convergence of parameter optimization;
(e) inability to discriminate between global and local optima.

Under this proposed framework, it should be emphasized that the need to better understand, jointly, the natural atmospheric and hydrological phenomena constitutes a compelling challenge that requires attention. Scientific questions arise to address water availability estimates uncertainty, scenario of extreme events forecasting, and global climate change. This case study aims to identify the dominant hydrological processes in order to establish monitoring guidelines to capture the dynamics and impact of natural hydrometeorological phenomena at the scale of the river basin.

Time series datasets regarding in situ measured variables were explored especially for precipitation, meteorological variables, and river flows together with estimates derived from remote sensing of precipitation, soil moisture, and spectral vegetation indices. The variables were analyzed on hourly, daily, monthly, and yearly scales. Spatiotemporal patterns of rainfall, soil moisture, and vegetation cover were delineated based on techniques of deterministic and stochastic analysis of spatiotemporal series, contributing to the improvement of the hydrologic assessment and use of rainfall-runoff models. The adopted procedures are applicable to predicting, forecasting, and monitoring of extreme events and to water resources management at the river basin scale.

2.1 Study Area

The Piabanha river basin has a drainage area of 2,065 km^2 and is located on the reverses of Serra do Mar (Fig. 1) in the state of Rio de Janeiro. It is worthwhile to note that the studied area is part of the Paraíba do Sul river basin that covers partially the states of São Paulo, Rio de Janeiro, and Minas Gerais in the southeast of Brazil, encompassing along its drainage area of approximately 55,000 km^2. Among some relevant features, it should be noticed that about 6.2 million people live in the referred basin. Approximately 10% of the Brazilian GDP is produced in the basin, namely around US$ 169 billion (Braga et al. 2008). Complementarily, the Paraíba do Sul river is the main source of water supply for more than 12 million people from Rio de Janeiro, including 85% of residents of the metropolitan area of Rio de Janeiro (MARJ) (ANA 2013). This diversity of human occupation leads to challenges that should be addressed with respect to land use and soil cover jointly with receiving sewage and industrial effluents sanitarily treated at different levels.

The cities of Petrópolis and Teresópolis are respectively placed on the headwaters of the Piabanha river and of its tributary Preto river. Both cities present some disordered occupation featuring accelerated growth. In rural areas, vegetables, some fruits, and flowers are grown, with inadequate agricultural practices and soil management due to high slopes of the relief, in addition to the uncontrolled use of pesticide and fertilizer as well as inefficient use of water by the adopted irrigation methods.

The hydrological processes are intensively monitored and studied since the soil parcels scale up to the systematic hydrometeorological monitoring in basins of different sizes. More specifically, one representative basin and three experimental basins of varying magnitudes and vegetation cover are available in the area.

The representative basin, with its outlet at the Pedro do Rio streamflow gauge station, has a surface area of 429 km^2 and contains three subsets of experimental watersheds, with different uses and land cover (LULC) (Fig. 1). In each subset, there are nested monitored drainage areas, ranging from 3 to 30 km^2.

New measuring instruments with different purposes have been continuously incorporated into the experimental watersheds. The water levels of rivers, their flow rates and water quality parameters are monitored, as well as precipitation, air temperature, air relative humidity, wind speed and direction, atmospheric pressure, incident solar radiation, net radiation, and soil moisture.

Since 2006, different institutions are involved in developing research work, including Federal University of Rio de Janeiro (Universidade Federal do Rio de Janeiro—UFRJ) by means of the Alberto Luiz Coimbra Institute for Postgraduate Studies and Research in Engineering—COPPE, Engineering Polytechnical School—POLI and Institute of Geosciences—IGEO, Mineral Resources Research Corporation (Companhia de Pesquisa de Recursos Minerais—CPRM) of the Brazilian Geological Survey (Serviço Geológico do Brasil—SGB), State University of Rio de Janeiro (Universidade do Estado do Rio de Janeiro—UERJ), State Environmental Institute—Rio de Janeiro (Instituto Estadual do Ambiente—INEA/RJ).

Later on, other departments and institutions became involved, such as the Institute of Biophysics, National Museum, and the Ecology Program, all from Federal University of Rio de Janeiro (UFRJ), Technological Centre of Mineral Resources (Centro de Tecnologia Mineral—CETEM), Federal Rural University of Rio de Janeiro (Universidade Federal Rural do Rio de Janeiro—UFRRJ), Federal University Fluminense (Universidade Federal Fluminense—UFF) and Brazilian Agricultural Research Corporation (Agriculture Empresa Brasileira de Pesquisa Agropecuária—EMBRAPA). In fact, several researches are opened to a diversity of collaborations and partnerships, noticing that the previously mentioned institutions are already active in developing joint projects in the region.

The area was chosen for its proximity to the various research centers, up to about 100 km away from any of them, and for presenting real challenging problems that are related to insufficient sanitation utilities, inadequate solid waste disposal, deficient urban drainage, intense erosion, high sediment load, industrial and diffuse pollution, all of them seriously affecting water quality. In addition, the region is subject to intense and frequent rains that cause floods and landslides, with numerous victims, as occurred in the well-known worldwide mountain mega-disaster in January 2011 and more recently in February and March 2022.

In December 2006, the Environmentally Integrated Basin Experiments (EIBEX)—Hydrological Parameterization started installing the monitoring network financially supported by the national public Call MCT/FINEP/CT-HIDRO—Representative Basins 04/2005, having as executor COPPE-UFRJ and collaborators IGEO/UFRJ-UERJ-CPRM-INEA. The basin with its established monitoring network has been conceived as a school-type basin. Indeed, this was the case, for instance, of the Hydrometry Extension Course (2008–2010) supported by the national public Call MCT/CNPq/CT-Hidro no. 037/2006, which trained technicians and students.

Its maintenance, since its implementation, has been guaranteed also by the institutional project of CPRM/SGB—EIBEX—Studies in Experimental and Representative Basins—with significant financial resources necessary for current expenses, in order to guarantee the monitoring network and its operation, with investments in new equipment for replacement and testing new technologies and training personnel. The support ensures continuous monitoring with field and office teams that collect and analyze data.

In the last twelve years (2010–2022), studies developed in the experimental basin were awarded new financial resources accordingly to the Call MCT/FINEP/CT-HIDRO 01/2010—Redes HIDROECO and scientific network so-called RHIMA, as well as from projects financed by the State Research Financing Agency—Rio de Janeiro (FAPERJ), namely Scientist of Our State, Thinking Rio and Emergency Research calls), National Council for Scientific and Technological Development (CNPq) funding for the so-called Universal projects and research grants provided by federal funding agencies such as CAPES, CNPq, and FINEP.

Complementarily, financial funding is provided by the Piabanha River Basin Committee (Comitê da Bacia do Piabanha—CBH Piabanha), established in 2006, which recognizes the importance of the research to support water resources management applied to basins in the mountainous region of Rio de Janeiro. The CBH Piabanha integrates the SINGREH and its available budget for investing in eligible technical or research projects is made possible by charging the use of water in the basin. Furthermore, the Piabanha River Basin Committee, which has its own statute and regiment, is articulated through an operational management contract with the Paraíba do Sul River Basin Agency (Agência de Bacia do Vale do Rio Paraíba do Sul—AGEVAP).

Once the legal framework has been established and an expanded overall view of the study area has been provided, it should be noted that, throughout the period of existence of the experimental basin project, there were, in the region, at least three extreme precipitation events with associated natural disasters—2008, 2011, known as the mountainous region mega disaster, 2013 and 2022, these last two more restricted to the city of Petrópolis. There was also the prolonged drought period along the 2013–2014–2015 time frame.

Extreme intense rainfall events are recurrent in the mountainous region of Rio de Janeiro State. The rainy season starts in September extending until April. The rainiest trimester is December–January–February, along which the mean precipitation amount is 45% of the mean annual total, i.e., 900 mm in three months out of the annual amount of 2.000 mm. In the lower lands, this pluvial distribution is yet more concentrated, the rainiest trimester sums up 50% of the annual total, i.e., 650 mm out of 1.300 mm. The driest trimester is June–July–August, as can be seen in Fig. 8.

The existence of the data generated by the project motivates and makes it possible to study hydrological processes that occur at different scales and still contributes to guide the solution of some real problems in the basin.

Digital Cartography

There are plenty of available cartographic digital maps for the municipality of Petrópolis in different scales from a cadastral scale of 1:2,000, for the urban area restricted to the region of the historic center, restored from aerial photographs at a 1:6,000 scale. For the other areas in the county, the available cartography has a scale of 1:10,000, resulting from the same aero-photogrammetric survey. In 2012, the Brazilian Institute of Geography and Statistics (IBGE) and INEA/RJ, in a joint initiative, made the topographic map available on a scale of 1:25,000 for the whole Rio de Janeiro state. There are also digital cartographic data bases at smaller scales, such as 1:100,000 (Fundação CIDE, 1994, without topography), and the topographic maps 1:250,000 (CPRM 2002) and 1,000,000 (IBGE, CPRM and ANEEL 2001) covering the entire basin.

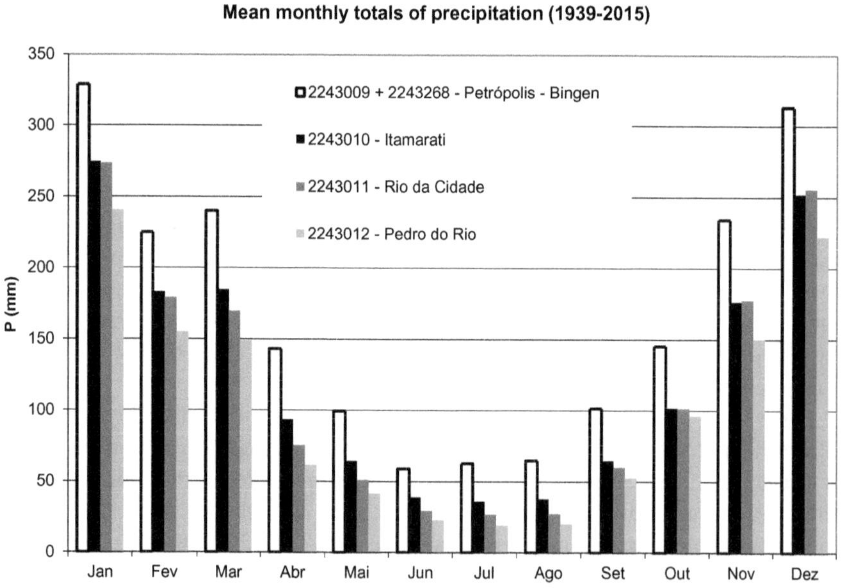

Fig. 8 Precipitation regime over the Piabanha river watershed

Digital Elevation Model (DEM)

Digital elevation models (MDEs) were generated by CPRM from the different scales of the available topographic maps, 1:10,000, 1:50,000, and 1:250,000, in addition to 1:25,000 generated by INEA/RJ. Due to the rugged relief in the region, the use of a coarser MDE to delineate the watershed and to extract physiographic characteristic indices for hydrological assessment has non-significant loss of information. The MDE in the 1:50,000 mapping scale was used for the purpose of delineating the watersheds and their morphometry as illustrated in Fig. 9.

Land Use and Land Cover Mapping

There are maps of land use and cover from different sources and scales. One of them is namely the map of the Ecological Economic Zoning of the State of Rio de Janeiro (ZEE-RJ 2008) at 1:100,000 scale. Another one is illustrated in Fig. 10, which was produced in the research developed by Rosário (2013) at 1:25,000 mapping scale.

Rosário (2013) developed a land use and land cover map for the Piabanha river basin to assess the impact of Brazilian law for native vegetation protection on the forest remnants of the Atlantic Forest biome. Images of the AVNIR-2 sensor from the 2009 ALOS mission were used with a spatial resolution of 10 m and orthophoto taken in 2006, compatible with the scale 1:25.000. Its legend is compatible with that of ZEE-RJ (2008) on a scale 1:100.000. Thus, both maps can be used together when information from outside the basin is necessary, as there are no discontinuities at the edges, only a reduction in detail. The distribution of the land use and land cover classes in the representative basin, according to the map of Rosário (2013), is discriminated in Table 1. It is interesting to note the predominance of the forest class in the basin (51.5%) and the expressive presence of rocky outcrops.

Population Distribution

The population distribution in the basin is depicted in Fig. 12. The features in the map are census sectors associated with data from the 2010 Demographic Census Universe (MPOG 2011). Points of water withdrawals and wastewater release (ANA 2012) are also shown in the map.

In Fig. 11, it is possible to perceive the great anthropic pressure on the ecosystem, as a consequence of the intense occupation in areas classified as conservation units. The mega disaster of January 2011 in the Piabanha river basin was certainly the consequence of highly intense rainfall over densely occupied areas.

2.2 Hydrological Modeling Using the Lumped Rainfall-Runoff Soil Moisture Accounting Procedure (SMAP) Model

This section illustrates the importance of hydrological modeling for a better understanding of the water and energy balance and the behavior of some hydrological

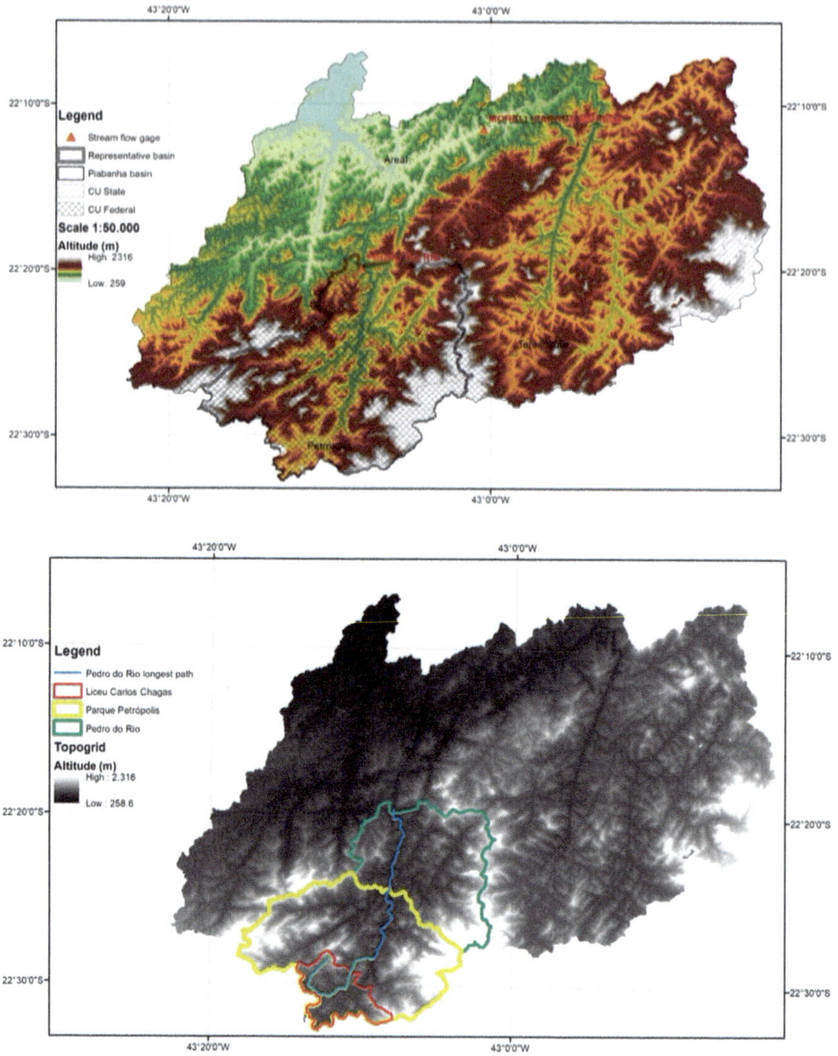

Fig. 9 Digital elevation model at 1:50.000 scale (Moreira et al. 2007)

variables in a basin. The conceptual lumped rainfall-runoff SMAP model (Lopes et al. 1982) was used. The SMAP model has as input data daily flows observed at the fluviometric station of interest, daily precipitation averaged within the basin from several rain gauge stations representative of the spatial variability of rainfall and potential daily evapotranspiration (ET0) or actual evaporation rate (ETR) estimates for the basin. The model structure is presented in Fig. 12.

The SMAP model was firstly coded in Fortran with optimization by hyperbolic smoothing procedure. The hyperbolic smoothing method (Xavier 2001, 2010; Xavier

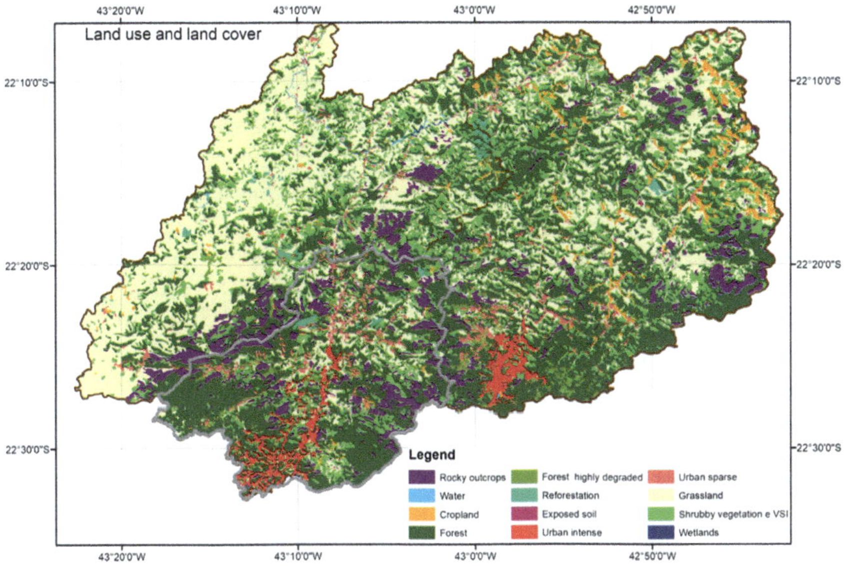

Fig. 10 Land use and land cover of the Piabanha river watershed (Rosário 2013)

Table 1 Land use and land cover within the study area (Rosário 2013)

Land use description	Area (m^2)	%	Land use description	Area (m^2)	%
Forest	221.557.198	51.5	Rocky outcrops	51.001.399	11.8
Forest (highly degraded)	8.531.498	2.0	Water	135.457	0.03
Reforestation	3.092.738	0.7	Cropland	1.756.264	0.4
Bare soil	392.326	0.1	Shrubby vegetation VSI	44.258.084	10.3
Densely urbanized	26.416.614	6.1	Grassland	52.110.593	12.1
Sparsely urbanized	21.298.913	4.9			

et al. 2005), implemented in the SMAP version, allowed to partially overcome some of the inconveniences related to the model's response surface (e.g., Johnston and Pilgrim 1976; Kitanidas and Bras 1980; Sorooshian and Gupta 1980; Sorooshian and Arfi 1982; Sorooshian et al. 1983; Gupta and Sorooshian 1985; Sorooshian and Gupta 1985; Duan et al. 1992; Yapo et al. 1998; Kavetski and Kuczera 2007).

This response surface is based, according to the implementation performed, on the objective function of least squares, which includes the observed flow and the simulated flow. The flow generated by the model is a function of the input information, of the states of the model's reservoirs, with the presentation of several water

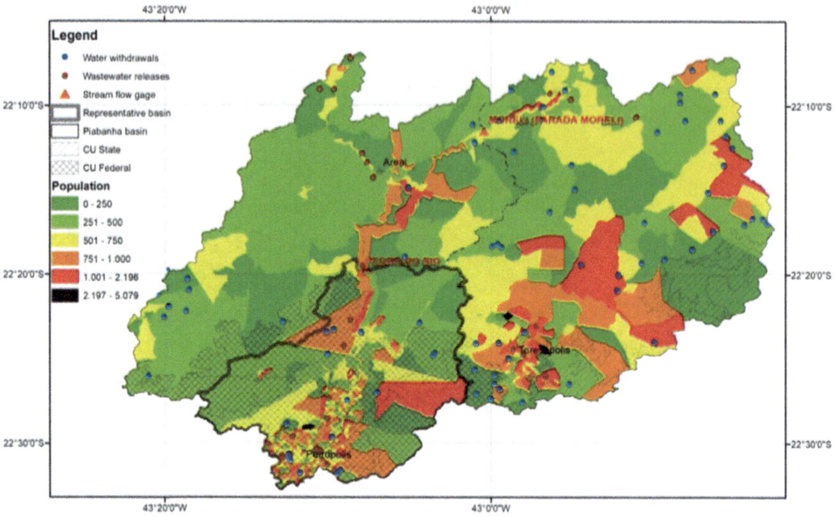

Fig. 11 Population distribution by census sectors featuring water abstraction and wastewater releases (MPOG 2011; ANA 2012)

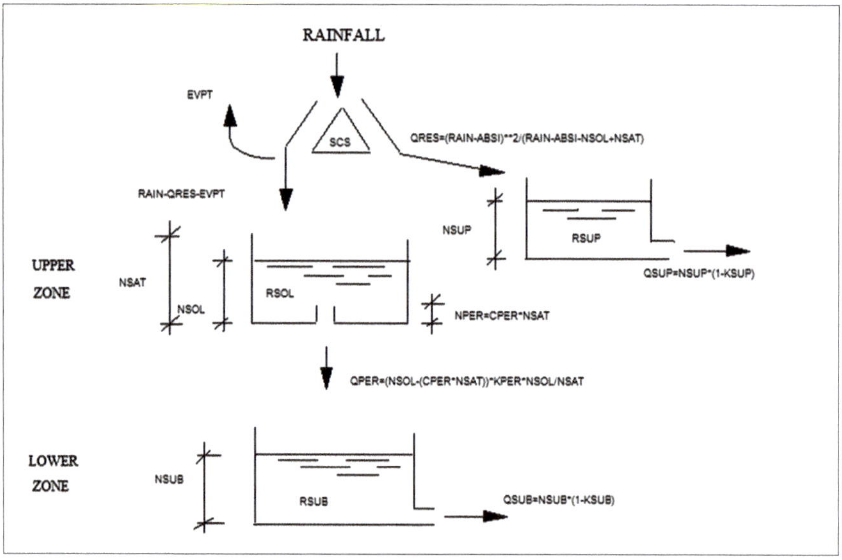

Fig. 12 Schematic representation of the conceptual rainfall-runoff model. SMAP with three reservoirs (Rotunno Filho 1989)

paths according to the model's computational algorithm, recursively in time. The smoothing method assumes rewriting the model's computational code in a different way from the traditional computational format of rain-flow-type physical–mathematical algorithms with reservoirs. It is noteworthy that the physical integrity of the model is maintained. Complementarily, when changing the response surface, by smoothing it, changes to the interactions in the n-dimensional space of model parameters are imposed, which explains the differences identified in the comparative analysis of the two versions of the model with respect to the optimal set of parameters obtained with the different optimization procedures implemented.

The implemented algorithm of the SMAP model checks if there is enough rain to supply the evapotranspiration, otherwise it consumes the soil moisture rate, checks the condition of the soil reservoir for percolation, and updates the contents of that reservoir. Thus, it is possible to provide the potential evapotranspiration, and the model itself calculates the actual evapotranspiration.

2.2.1 Hydrological Simulation with Model SMAP at Pedro Do Rio Streamflow Gauge

An optimal set of parameters of the SMAP applied to Pedro do Rio was obtained by Gonçalves (2008), who used the version with the hyperbolic smoothing method, as used in Di Bello (2005) and presented in Rotunno et al. (2014). Further details can be found in Xavier et al. (2022). The model was then calibrated and validated, respectively, for the periods 1998–2001 and 2001–2004.

Runoff (surface flow) was overestimated along the annual recessions during both periods in comparison to observed streamflow values, having the calculated mean flow of the period resulted 18% higher than the observed one. In agreement with Gonçalves (2008) concluding remarks, the daily time step of the simulation was also judged to be inadequate, being much larger than the time of concentration of the watershed (8 h). The available observed flow time series was being registered at that time as daily means at Pedro do Rio streamflow gauge. In 2016, a continuous recording gage was installed at the station.

For exploring the meaning and effectiveness of the parameters and the structure of the SMAP model, a version implemented in the format of a spreadsheet, with optimization using the solver function, was firstly run with the same set of optimal parameters, found by Gonçalves (2008) for two time periods, namely 1998–2001 (calibration) and 2001–2004 (validation), in order to compare the algorithms of the two versions. The spreadsheet model achieved almost the same performance, which was verified by the following adopted metrics: Nash–Sutcliffe coefficient, coefficient of determination and root mean square error (RMSE) (Figs. 13 and 14). The differences could be attributed to slightly different internal approximations in terms of the implementation of the algorithm or to some interaction between parameters captured differently by the hyperbolic smoothing method.

Based on experiments and collected data in the basin, some of the SMAP parameters which are supposed to have physical meaning, were estimated and tried in a new

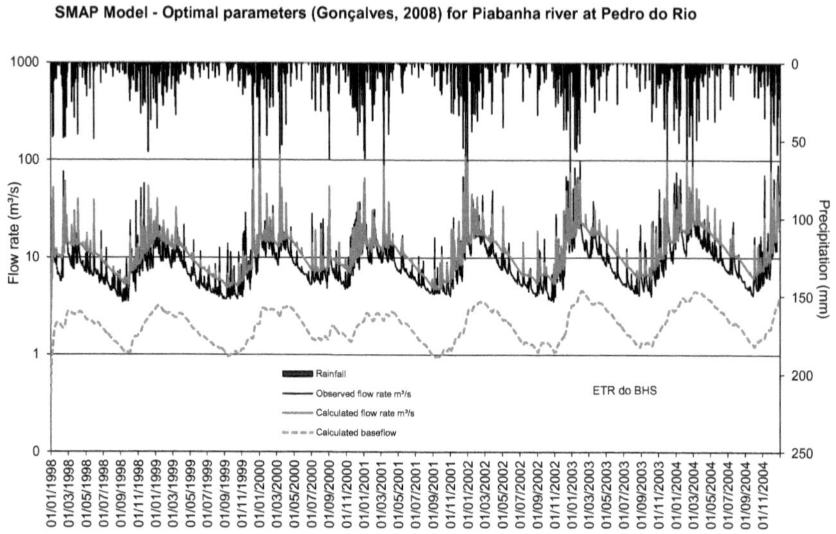

Fig. 13 Observed and simulated streamflows along 1998–2004 (calibration and validation periods) adopting the set of optimal parameters stated in Gonçalves (2008)

simulation. The evapotranspiration was estimated using observed data from the two weather stations located in the basin and based on two different methods as Camargo (1971) and Penman–Monteith (Allen et al. 1998). The results for the same period (1998–2004) were quite different as the overestimation was reduced along the recessions, with underestimation of high flows (Figs. 15 and 16), having the calculated mean flow of the period resulted 8% lower than the observed one.

Through the analysis of all the intermediate results attempting to find the best set of parameters for simulating the whole period (1994–2014), the conclusion was that, besides the inadequate time step of the model, other reasons could be contributing to not so good results, The interflow might be significant in the basin, and the structure of the SMAP does not represent this reservoir. Other models could be more suitable to this basin, however the model is quite robust for supporting decision-making on water resources management issues that require short time response.

2.2.2 Discussion—Basics on Hydrological Modeling

The problem of whether to employ a simpler model or a complex rainfall-runoff model to support a decision-making on water resources issues, is really just a smaller facet of a larger and challenging problem on how to decide on which level of complexity the model structure should be established to appropriately represent the physical processes taking place at a given watershed. At the heart of this problem is the issue of system scale, which is often ignored or scarcely treated.

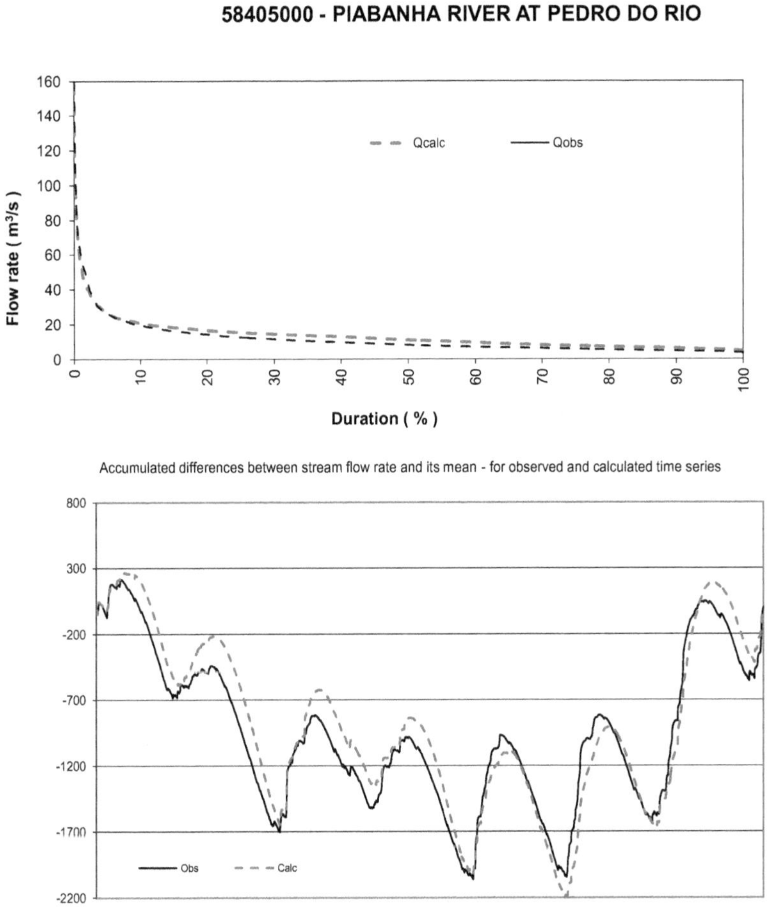

Fig. 14 Observed and simulated streamflows along 1998–2004 adopting the set of optimal parameters found by Gonçalves (2008)—streamflow curves (top); accumulated differences with respect to the mean flow (bottom)

For instance, several analyses have been carried out about uncertainty in calibration data and in parameter fitting of rainfall-runoff models such as in Troutman (1982, 1985a, b) and in Kuczera (1983), among others. Troutman (1982, 1985a, b) investigates the errors related to the input variables and how these errors affect the output variables. The author provides a framework for theoretical studies to be conducted about the structure of errors in rainfall-runoff models. Kuczera (1983) develops a bayesian methodology to evaluate parameter uncertainty in catchment models fitted to a hydrologic response such as runoff, the goal being to improve the chance of a successful regionalization. The methodology proposed by Kuczera (1983) considers how to combine several kinds of hydrologic data to improve the precision of poorly determined parameters and also how to test their compatibility.

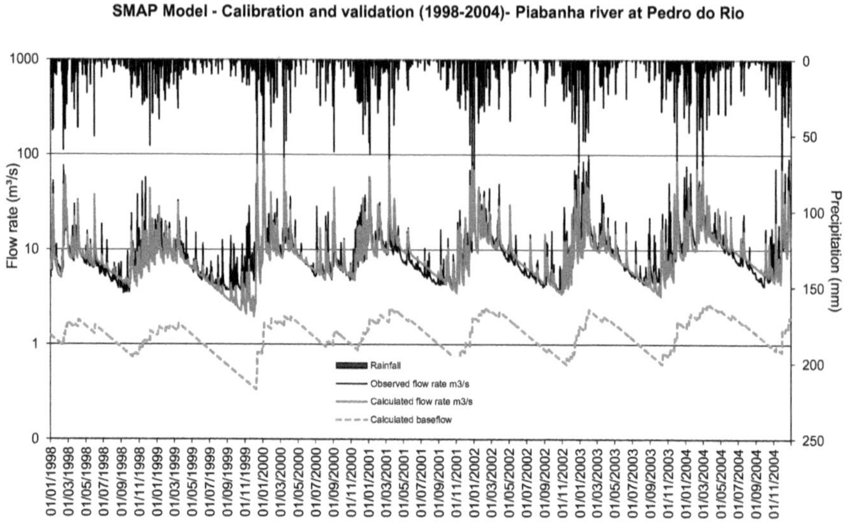

Fig. 15 Observed and simulated streamflow rates along 1998–2004 adopting parameters based on soil characteristics, streamflow recession mean behavior and ET0 estimated for the watershed

Furthermore, Klemes (1983, 1986), Beven and Binley (1992), and Jensen and Montaglou (1992) pose several questions of how one might move from lumped model to semi-distributed and distributed models taking into account uncertainties and conducting proper calibration and validation procedures questioning some basics on hydrological principles. Such issues are on-going challenges. Baldassarre et al. (2013) and Troy et al. (2015) rephrase the hydrological agenda emphasizing the term socio-hydrology since the hydrological basin is indeed an integrated physical and social system. More recently, Blöschl et al. (2019) also made an effort to establish a scientific agenda based on what the authors called 23 unsolved problems of hydrology based on an analogy to the 23 unsolved problems in mathematics proposed by Hilbert (1902).

Accordingly to Beven and Freer (2001), the Generalized Likelihood Uncertainty Estimation (GLUE) methodology offers an alternative to estimate the uncertainty in hydrological model output, by selecting a set of simulators of the watershed response that can be accepted as behavioral. The concept of behavioral simulators originates in the paradigm of equifinality. Due to the inherent difficulties concerning the search of an optimal parameter set of a model, it is often only possible to find a number of parameter sets which simulate the system behavior equally well according to a quantitative measure of goodness of fit that must be defined a priori (Beven 1989).

Similarly, Xavier et al. (2005) explored the GLUE framework for the estimation of the total uncertainty of the hydrological model TOPMODEL (Beven et al. 1995) that comes from the approximation of a precipitation field from different layouts of a rain gauge network. Given a number of rain gauge stations and their time series, the precipitation field can be approximated using a geostatistical approach through the

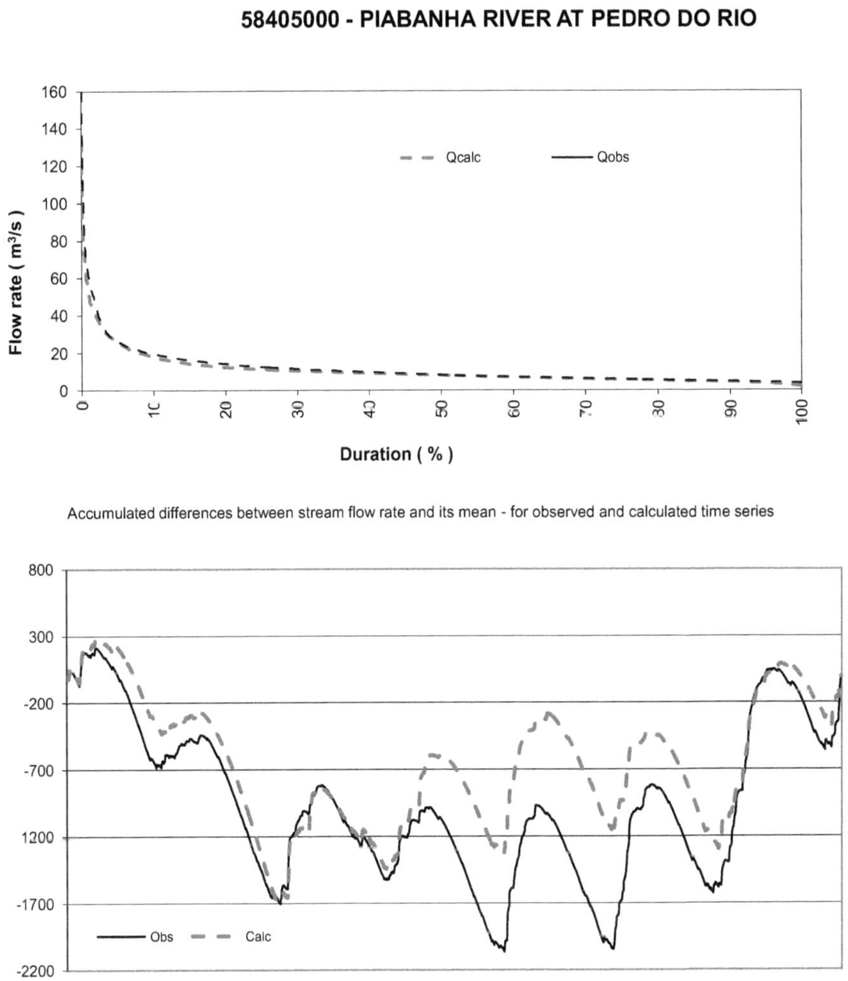

Fig. 16 Observed and simulated streamflows along 1998–2004 adopting parameters based on soil characteristics, streamflow recession mean behavior and ET0 estimated for the watershed—streamflow duration curves (top) and accumulated differences with respect to the mean flow (bottom)

Bayesian Kriging methodology, which combines the point measurements given by rain gauges with the knowledge of the spatial precipitation structure, which can be estimated with remotely sensed images (e.g., optical and radar images). It could be expected that the greater the number of these rain gauges, the better the approximation of the real precipitation field would be, and thus the less significant the uncertainty contribution caused by the imperfect knowledge of the real precipitation field.

In addition, recent scientific initiatives focus on the development of methodologies using random multiplicative scaling to increase spatial resolution of satellite information, such as the remotely sensed precipitation. The purpose is to quantify precipitation statistics over a wide range of scales and in a range of process dynamics. For instance, Vásquez (2018) adopted multifractal and waveforms modeling approaches for addressing downscaling analysis of the space–time structure and rainfall based on initial estimates retrieved from Tropical Rainfall Measurement Mission (TRMM) satellite with 0.25 degree (approximately 25 km) spatial resolution product. Successful results at the monthly and annual temporal scale were achieved up to the scale of approximately 3-km spatial resolution providing a good description of the mean and extreme events when compared to previously evaluated datasets at the 25-km spatial resolution.

It is noteworthy to say that hydrological simulation models have become an integral part of almost all aspects of hydrology. For example, better soil physical system and hydropedological descriptions (Ottoni et al. 2018, 2019), soil moisture and soil storage using observed and remotely sensed information (Rotunno et al. 2014), current and forecasted weather conditions (Silva et al. 2018, 2019), extreme events of floods (Souza et al. 2020) and droughts (Rebello et al. 2016) jointly with projection and simulation of monthly climatic scenarios under the guidelines established by IPCC (2007, 2013, 2016) (Luiz Silva et al. 2020) are inputs to models to estimate future flood discharges for emergency operations, flood and drought control, reservoir regulation and reservoir management (Luiz Silva et al. 2018; Dafne Chan et al. 2019; Treistman et al. 2020).

Models are used in planning studies to evaluate the benefits and impacts of alternative land uses, flood and drought control measures and also to develop soil services conservation practices as strategies to control soil erosion and land degradation in order to environmentally properly sustain the hydrological and associated biogeochemical cycles under the positive nexus water or water security, energy and food production. In design studies, models are used to assist in developing specifications for engineering structures to solve specific flood control, navigation, or water supply problems.

Under this broad and brief overview about basics of hydrological modeling, applications, and corresponding discussion about errors and uncertainties, an approach is then taken here-in envisaging to be further coupled to rainfall-runoff modeling in future studies at the watershed scale in order to explore hydrological similarities and patterns for identifying more reliable parameterizations for representing the flows based on the water and energy balance concepts, cycles and trends. Therefore, hereafter remotely sensed imagery is explored jointly with hydrological variables for identifying hydrological patterns using spectral analysis.

2.3 *Hydrological Patterns in the Piabanha River Basin*

Watershed physiographic characteristics and land use and vegetation cover are known to be related to spatio-temporal hydrological behavior. Pattern recognition techniques applied to grouping such characteristics offer the possibility to reveal potential relationships with hydrological variables and thus constitute an important approach to be applied in water resources assessment.

Investigation of connections between precipitation and climatic indices contributes to the predictability of severe rainfall events in a basin or region. Some meteorological systems could be early detected and operational forecasters could benefit to properly warn decision-makers on how to be prepared within the framework designed to adequately handle risk hazardous scenarios and to face an eventual critical situation.

Specifically, a detailed case study is developed with a greater focus on the spatio-temporal variability of vegetation and precipitation. The applied techniques could be extended to other hydrological and environmental variables of interest.

2.3.1 Database

Satellite products of different spatial resolution are available for estimating vegetation spectral indices. The chosen product for this analysis is the bimonthly normalized difference vegetation index (NDVI) composite derived from channels 1 (near infrared range wavelength) and 2 (red range wavelength) of the advanced very high resolution radiometer (AVHRR) on board the National and Oceanic and Atmospheric Administration (NOAA) Polar-orbiting Operational Environmental Satellites (POES). The data is described in Tucker (2005), consisting of a 8-km pixel NDVI time series retrieved for the period from 1981 to 2006 (26 years), downloaded from the Global Inventory Modeling and Mapping Studies (NASA-GIMMS 2014), hereafter called simply by GIMMS-NDVI. In order to cover the representative watershed, 25 pixels were considered in the analysis just to form a regular shape (Fig. 17).

Tropical Rainfall Measuring Mission (TRMM) satellite-based daily precipitation estimates, relative to 3B42RT product (NASA 2015a, b, c, d) and corresponding to pixel size 0.25° × 0.25° (~27.8 km × 27.8 km), are used in comparison with ground gauged data. The TRMM pixel centered at (−43.125; −22.375) practically covers the study area, the representative watershed.

Most of the daily data in Brazil is read from analogic conventional gauges at 7:00 a.m local time (UTC-3) and is assigned to the day of the reading. Thus, for comparison on daily basis, the 3 h TRMM estimates were summed up to every 9:00 a.m (UTC) and assigned to the same day. This procedure results in a lag of 1 h between data from the two sources, being observed data acquired 1 h later than the TRMM estimate. The hydrometeorological monitoring network is operated by the Geological Survey of Brazil (CPRM) jointly with the Brazilian National Water Agency obtained from

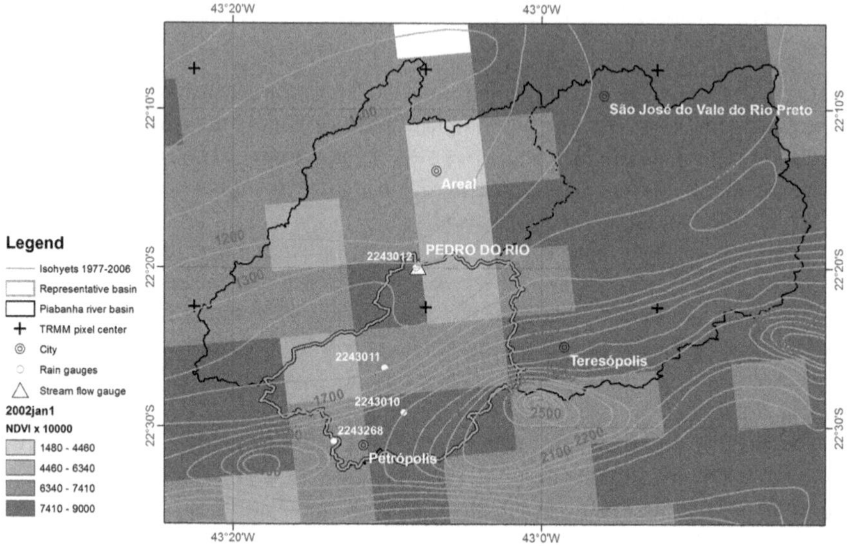

Fig. 17 Data sources used in the analysis—NDVI and TRMM pixels, four rain gauge stations and one streamflow gauge. The NDVI image was acquired on January 1, 2002

(ANA). Precipitation amounts time series gauged at Petropolis-Bingen (2243009-268), Itamaraty (2243010), Rio da Cidade (2243011) and Pedro do Rio (2243012) stations were downloaded from ANA (2015).

Gauged precipitation and satellite estimates were compared along the period 2000–2014. The comparative evaluation is shown in Fig. 18 where time series statistical metrics of annual rainfall water amounts are presented in the form of boxplots. Areal satellite estimates for the central TRMM pixel in the basin and point observed precipitation show very close temporal distribution, especially similar to Rio da Cidade rain gauge station (22243011), but not for Petropolis-Bingen station (2243009-268), which is located in the upper part of the watershed.

The available dataset and adopted time frames comprehending all analyzed variables can be summarized as follows:

(i) data source GIMMS-NDVI -NOAA/AVHRR, time step 15 days, available period 1981–2006, analyzed periods 1981–2006 and 2001–2003;
(ii) data source TRMM 3B42RT product (3 h), time step 3 h, available period 2000–2016, analyzed period 2001–2003;
(iii) data source rain gauges CPRM/ANA 2243009–288, 2243010, 2243011, 2243012, time step 24 h, available period 1939–2016, analyzed periods 1939–2014 and 2001–2003.

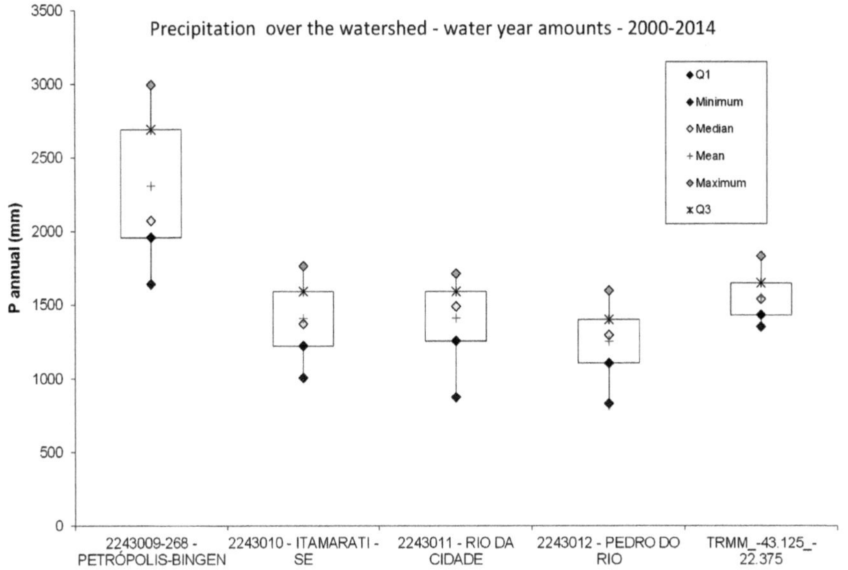

Fig. 18 Statistical metrics for annual amounts of precipitation (2000–2014) collected at three stations and retrieved from satellite estimates at four pixels. The pixel covering most of the watershed (centered at −43.125°; −22.375) and the central rain gauge station (Rio da Cidade) are highlighted for comparison. Petrópolis-Bingen gauge station is located in the upper part of the watershed where precipitation is not adequately estimated by the TRMM satellite sensor

2.3.2 Discrete Wavelet Transform (DWT) and Multiresolution Analysis (MRA)

For a joint evaluation of precipitation, streamflow, and NDVI for vegetation cover dynamics in the representative basin along the 2001–2003 time span, all involved time series were analyzed with the Haar (1910) discrete wavelet transform (DWT) followed by a multiresolution analysis (MRA).

Vásquez et al. (2017), using DWT and MRA, for gauged precipitation (2243010, 2243011 and 2243012), TRMM estimated (at −043.125; −22.375) and streamflow at Pedro do Rio daily time series along 2001–2003, reached the decomposition in a progression of approximations and successive details in the different time scales 2^{-j}, for j varying from 1 to 6 and $\Delta t = 1$ day, with D1 = 2 days, D2 = 4 days, D3 = 8 days, D4 = 16 days, D5 = 32 days and D6 = 64 days.

Comparing the variability of these details, the conclusion was that 2243011 (Rio da Cidade) was the most representative rain gauge compared with streamflow responses at Pedro do Rio. Compared with TRMM estimated precipitation at the pixel centered in the basin, the most similar rainfall time series in terms of its variability was also the one gauged at 2243011. The similarity was greater for longer accumulation period, expressed by the variability of the details. The central TRMM pixel in the basin represents the average precipitation well, which could be seen in the comparison of the temporal variability of this series with that of the streamflow series at Pedro do Rio. Accumulating the amounts of precipitation over stations 2243011 and 2243010 for every fortnight along 2001–2003, the resulting time series were analyzed with the same DWT-MRA approach jointly with GIMMS-NDVI time series of pixels with centers at—(43.151; −22.501) and (−43.159; −22.431), The results showed the same temporal variability for NDVI and precipitation in the time intervals of the details: D1 = 1 month, D2 = 2 months and D3 = 3 months, considering $\Delta t =$ 15 days of the original series.

2.3.3 Rescaled Range (R/S) Analysis and Hurst Coefficient Estimation

The R/S technique is applied to NDVI time series in order to investigate its dynamics, whether its variation has short or long memory, i.e., if it presents autocorrelation.

The technique was first proposed by Hurst (1951) to analyze the Nile river annual streamflow rates as inflows to evaluate the required reservoir capacity in the design of the Aswan dam.

It consists of dividing the time series into segments of different lengths (τi); then, for each segment, the mean is calculated, as well as the accumulated departures from the mean, the range between the maximum and the minimum accumulated departures from the mean (Ri), and the standard deviation (Si). In the sequence, the range is divided by the standard deviation (Ri/Si). Once this procedure is performed for every segment, the next step is to plot the natural logarithm of each Ri/Si against the natural logarithm of τi. Finally, a linear statistical regression is applied to obtain the slope of the adjusted line, which provides the correlation coefficient used to evaluate the quality of the fitting equation.

Hurst (1956) extended the analysis to other natural time series such as precipitation, temperature, pressure, sunspot numbers, lake levels and long-term records of tree rings growth and mud layers in lakes. He concluded that all the time series presented a similar slope coefficient in the linear fitting, all greater than 0.5, which is the expected value for time series of random variables, normally distributed with a finite population variance. His results were attributed to the tendency of natural phenomena to have runs of high or low values. The author concluded that some of the time series could complement datasets from river discharge in a long-term analysis, as in some way they are related to floods and have the hypothetically potential persistent behavior.

Many stochastic models were since that time developed to reproduce this effect of long-term memory on annual flows, which became known as Hurst phenomenon. The name of this effect was attributed by Mandelbrot and Wallis (1969), as "Hurst's law," since they referred to Hurst's results and partially confirmed his findings, stating that geophysical records would present global statistical dependency. They also labeled the slope coefficient of the adjusted line in the R/S technique with letter H, instead of letter K used by Hurst, then expressed as the exponent in the relationship $R/S \sim (\text{lag})^H$, where $H \neq 0.5$ for geophysical records exhibiting considerable global statistical dependence.

In complement, Rao and Bhattacharya (1999) tested two alternative methods to estimate the Hurst exponent and they found a robust behavior of the R/S and therefore they recommended its use.

On the other hand, Koutsoyiannis (2003) investigated the variability and the presence of Hurst phenomenon in long hydroclimatic time series. He concluded that Hurst coefficient estimate based on the rescaled range algorithm was inappropriate, as it depicted great dispersion of the results, a negative bias, and mathematically inconsistent values.

Katsev and L'Heureux (2003) highlighted some caveats to be avoided when analyzing time series shorter than 500 points or with large-magnitude abrupt changes in the variable of interest showing that it could lead to spurious values in estimating the Hurst exponent and also in misleading detection of fractal scaling.

It is also worthwhile to mention the work conducted by Chamoli et al. (2007), who tested five different techniques to estimate the Hurst coefficient, such as the wavelet transform and the method of rescaled range, which they concluded to be both adequate to estimate H for short or long data sets.

Complementarily, Villarini et al. (2011) estimated the Hurst exponent H to evaluate long-term persistence of annual maximum daily rainfall using the aggregated variance method. They attributed the uncertainty in their results partly due to the limited length of the time series.

Finally, it is interesting to remark that the Hurst effect is also explored as a technique to evaluate phenomena investigated in other scientific fields that present self-similarity, such as in cardiology, as applied by Julián et al. (2013) in analyzing cardiac arrhythmia, in fact, a generalization of the original approach by Hurst (1951).

The technique applied to the study region to evaluate its vegetation cover dynamics

The GIMMS-NDVI bimonthly time series was used as a proxy for vegetation cover, along the period 1981–2006, which consists of 612 NDVI values for each of the 25 pixels covering the study area. The R/S analysis was applied to each pixel time series associated with this corresponding length dividing it into different subsets of varying lengths from 612 to a minimum of 25 points. The result is the possibility of fitting of a linear relationship to 12 pairs of log (R/S) $\times$ Log (τ) values. The slope of this line is the Hurst coefficient or exponent, as the example presented in Fig. 19. For the R/S calculations, a MATLAB script initially coded by Enriquez (2014) was adapted accordingly.

The results presented in Fig. 20 revealed some differences in behavior for part of the watershed. Four NDVI pixels had Hurst coefficient (H) < 0.5, therefore considered non-persistent. These pixels are shown in lighter color in Fig. 20. For the other 21 NDVI pixels, the results show $H > 0.5$, indicating persistency.

If only the period 2001–2003 was considered in the analysis, all 25 NDVI pixels time series would be considered persistent as $H > 0.5$ for all of them (not shown). This is possibly a consequence of the short period of the analysis and since these three years denote wet time period, i.e., above the long-term mean, when vegetation cover responded positively in vitality to the high amounts of precipitation.

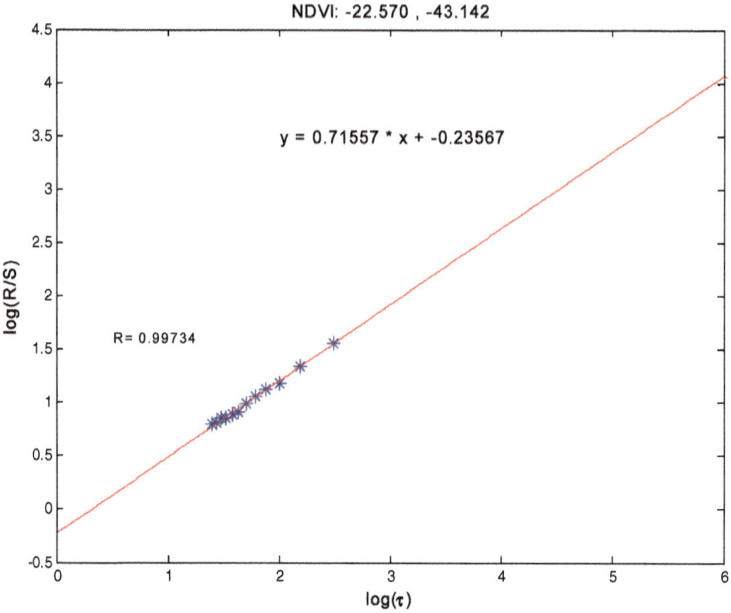

Fig. 19 Hurst coefficient definition applying the R/S technique. The example shows a persistent NDVI pixel centered at (043.142° W; 22.570° S) as the slope coefficient resulted greater than 0.5, i.e., $H = 0.71557$

It is seen in Fig. 20 (top) that the persistent pixels are mostly distributed along areas where forested and consolidated urban features predominate. The four non-persistent pixels are distributed along grassland, near reforestation sites or where forest cover is quite fragmented.

A simplified hydrogeological map, which corresponds to 1:1.000.000 scale (CPRM 2014), is also used as support to present spatial distribution of the results (Fig. 20—bottom). Non-persistent NDVI pixels are located on hydrogeological formation (4) Fr—Fractured unity, generally associated with low productivity, although locally moderate, different from the rest of the watershed that is located on (6) Fr—Fractured unity, low productivity, not aquifer (CPRM 2014). The area where the non-persistency occurred is also a less rainy region, where the annual amounts are expected to vary from 1200 to 1300 mm.

The non-persistent pixels are located in the Fagundes watershed, considered not to properly belong to the homogeneous region defined for Piabanha river basin in previous studies on stream flow regional analysis (CPRM 2002). Differences in physical characteristics between the areas suggest that the Hurst coefficient could be a fair and good indicator for recognizing NDVI as an important parameter for characterizing spatiotemporal variability of vegetation and soil cover and corresponding hydrological behavior patterns.

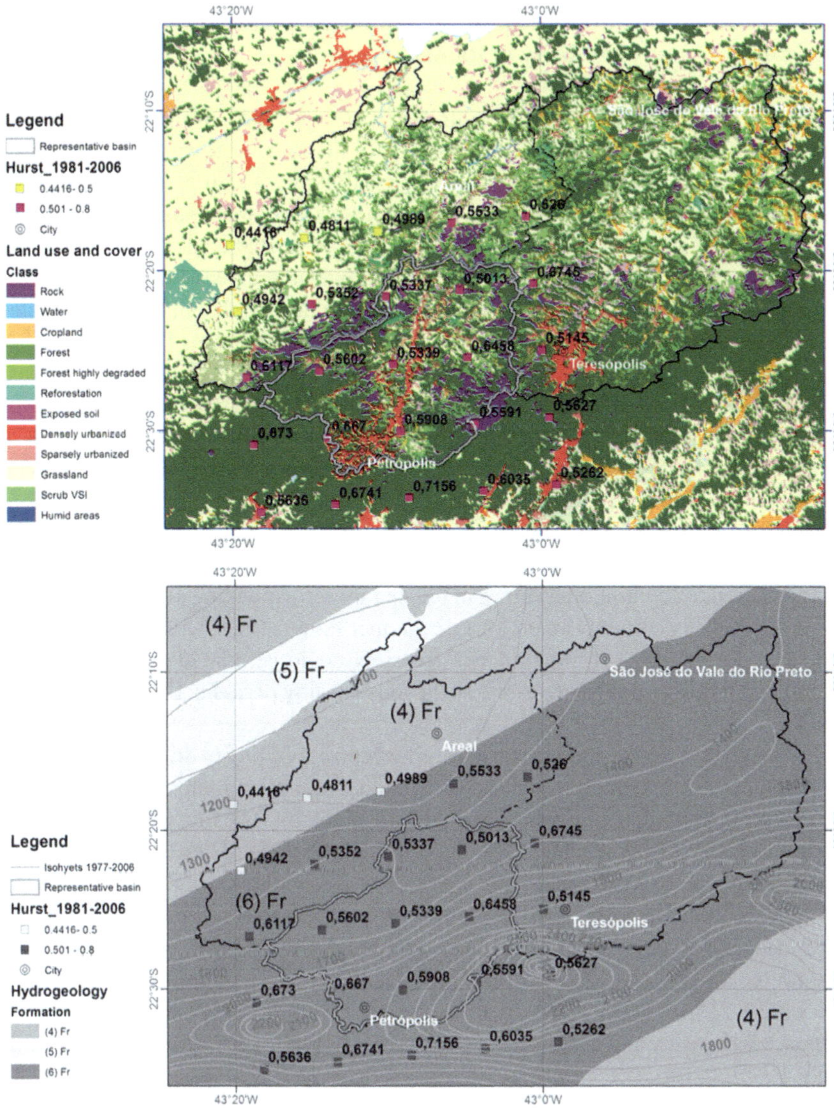

Fig. 20 Results for NDVI pixels in R/S analysis for the representative basin: land use and vegetation cover map (Rosário 2013) (top panel)—four pixels had non-persistent $H < 0.5$ on grassland and 21 pixels had $H > 0.5$ on forest; hydrogeological map (bottom panel)—formation where NDVI pixels show persistence is (6) Fr and formation where the pixels show non-persistent is (4) Fr (CPRM 2014). Annual precipitation is approximately 1,300 mm where NDVI pixels are considered non-persistent, and it is above 1,400 mm where NDVI resulted persistent

2.3.4 Continuous Wavelet Transform and Spectrum Analysis to Evaluate Periodicity in Rainfall Features

In order to identify periodicities in rainfall over the watershed, the available monthly longer time series, corresponding to three gauge stations, were analyzed by means of the continuous wavelet transform (CWT) technique, which is an adequate tool for localizing intermittent oscillations in a time series. It was applied by means of using the MATLAB software package provided by Grinsted et al. (2004) following Torrence and Compo (1998), who firstly made available an easy-to-use toolkit for wavelet analysis that included statistical significance testing. Labat (2005a, b) states that CWT improves determination of the temporal-scale structure of a given signal as it overcomes the well-known limits of the classical Fourier analysis, that is restricted to the frequency domain of stationary periodic processes, which is not usually the case of hydrometeorological datasets. For further details, fundamentals and principles of the theory behind the technique are clearly presented in the three above-mentioned references and well complemented in Daubechies (1992).

The results for the three 76-year monthly precipitation time series presented very little variation among them (Fig. 21). The strong global power 1 year (12-month) period, an almost continuous feature in red and its corresponding peak on the power spectral graph, is certainly related to the annual rainy season with a maximum in January. The weaker 3–4 month period possibly could be related to seasonality of rainfall distribution along trimesters, as the mostly rainy period is DJF and the driest JJA.

Segmentations in the spectra were identified corresponding to drier periods of two to three years long, depicting accumulated annual rainfall amounts below the long-term mean in more than one and a half standard deviation. The first time period takes place around 1952–1954 (160 to 180 months after 1939) for 2243010 and 2243011, while the second one is around 1969–1971 (360–385 months) for the three stations, but more expressive for 2243012. There is a third one that can be noticed around 1983–1984 (540 months) and still another one around 1989–1990 (620 months), also for the three stations. For station 2243012, there is a discontinuity around 1957–1958 (220–225 months). To assist in the identification of these anomalies, dashed rectangles have been added to the figures involving the corresponding periods in Fig. 21.

The anomalies may also be identified by the standardized annual precipitation, calculated for the water year, along the whole period (1939–2015) presented in Fig. 22.

CWT was also applied to the monthly series of sea surface temperature (SST) of the Pacific Ocean, averaged in the region of the Niño 3.4 index—SST weighted within the area over the Pacific of 5° S–5° N and 170°–120° W (NOAA-ESRL 2015a, b).

In the spectrum of Niño 3.4, there were segmentations in the 12-month cycle, corresponding to the cold phase anomalies. An interruption of the period of 1 year around 1988–1989 would correspond to the La Niña (LN) in these two consecutive years. A significant contour of anomalies in the range of 2.7 (32 months) to 5.3 years

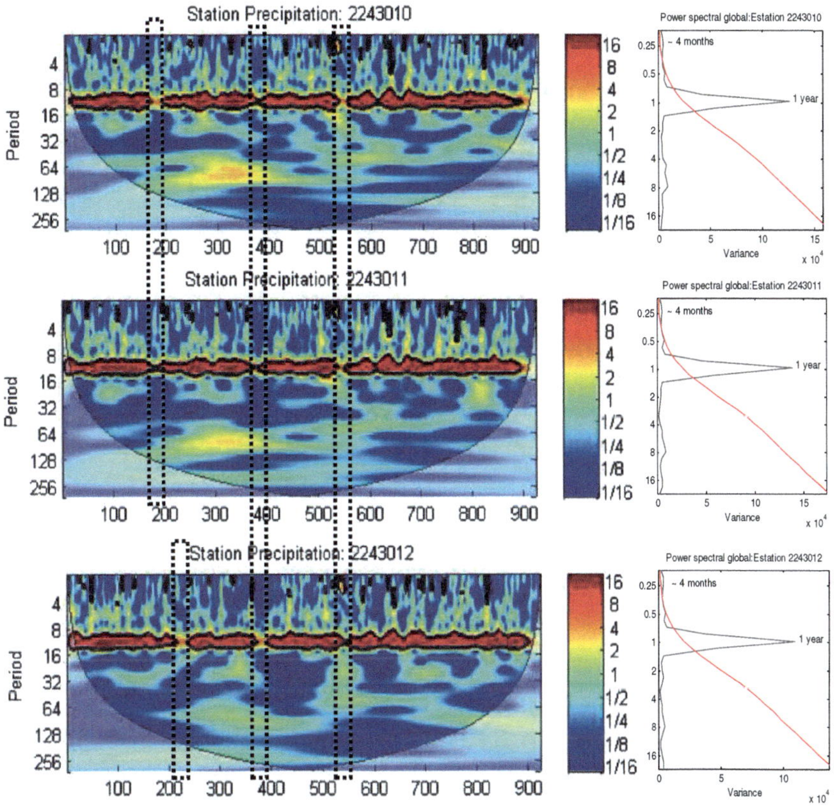

Fig. 21 Wavelet power spectrum (based on the Morlet wavelet) of 76 years monthly precipitation time series at gauge stations 2243010 (top panel), 2243011 (middle panel), and 2243012 (bottom panel) from 1939 to 2014. Periods (y-axis) and time (x-axis) are expressed in number of months. Global power spectra are shown in years

(64 months) of periodicity marks the periods of episodes of El Niño (EN) and LN, such as the EN of 1965–1966, the EN of 1972–1973, the 1982–1983 EN, the 1988–1989 LN, the 1997–1998 EN and the 2009–2010 EN.

To analyze the relationship between the precipitation and Niño 3.4—SST time series and if the phenomena have been in phase, the cross wavelet transformation technique (XWT) was applied to the TCW pairs of series of precipitation and Niño 3.4—SST, and then the wavelet coherence (WTC) applied to qualitatively assess the temporal evolution of the degree of linearity of the relationship between the two signals on a given scale. The whole procedure was repeated for all three stations, although the result is shown only for station 2243011, which is the most central in the basin (Fig. 23).

The XWT result is shown in an image (Fig. 23a) where the color scale indicates the common power in the time periods. The phase or antiphase in the time–frequency

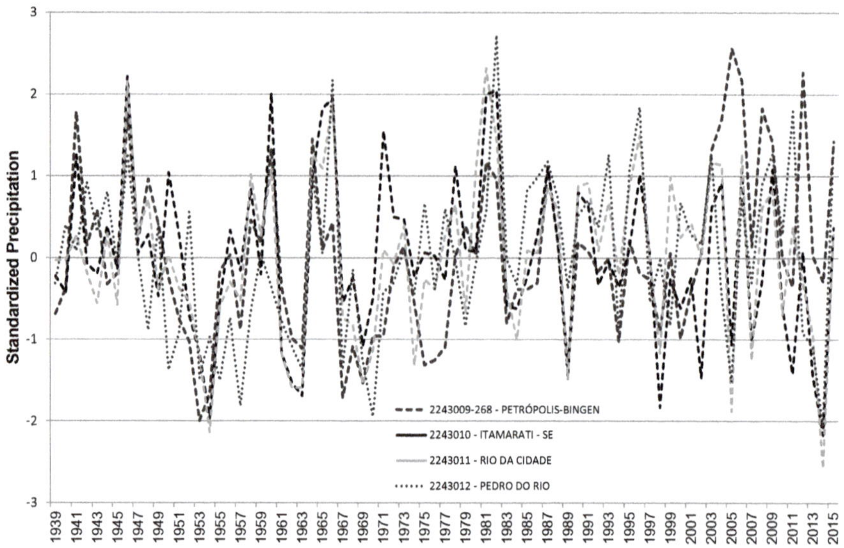

Fig. 22 Standardized annual precipitation amounts (departures from the mean divided by the standard deviation) showing the tendency of consecutive two to three-year duration of wet or dry periods

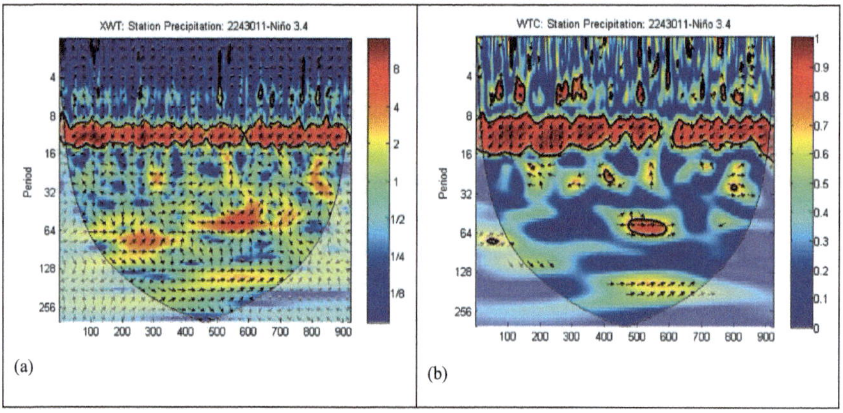

Fig. 23 **a** XWT and **b** WTC for precipitation over 2243011 and SST in the Niño 3.4 region monthly time series along 1939–2015

space between the two time series is indicated by the orientation of the arrows: to the right when in phase and to the left when in antiphase, the more horizontal the arrows are, the more strongly in phase or antiphase they would be. The correlation region with a significance level of 5% against red noise is shown within the black outline. Also shown is the cone of influence of the border effects, a light shaded area on both sides of the figure (Grinsted et al. 2004).

As in a correlation, the coherence between two series is greater the closer to 1 (Fig. 23b). It is possible to think of the WTC as a correlation coefficient located in the time–frequency space. Likewise, the orientation of the arrows indicates phase and antiphase, and the black outlines of the regions of the correlation indicate the significance level of 5% against red noise. In the representation of the WTC (Fig. 23b), the period shown, lasting from 500 to 600 months after 1939, in the range of variability from 48 to 64 months, presents the phenomena in phase. Its extension covers the 1982–1983 EN event, in which rainfall occurred above average in the basin, with positive SST anomaly and positive precipitation, and the 1988 and 1989 LN event, in which rainfall occurred below average in the basin, with negative SST anomaly and negative precipitation.

2.3.5 Discussion on Hydrological Patterns

The study region is a humid mountainous watershed in Rio de Janeiro state, where the mean annual precipitation amount varies from 1300 mm up to 3200 mm. In summary, it is possible to state that precipitation amounts over the watershed are expressive and rather regular. Even in the drier periods, annual totals never fell below 820 mm in the lower part of the basin (Pedro do Rio) or 1600 mm in its headwaters (Petropolis-Bingen), amounts associated with the driest water year 2014–2015 on the records. The long-term annual mean is higher than 1400 mm over most of the basin and 1300 mm at Pedro do Rio gauge station. For more recent periods, the exposed spatial distribution is confirmed by TRMM estimates.

The South Atlantic Convergence Zone (SACZ), which is an episodically persistent cloudiness and precipitation band NW–SE oriented, from the Amazonia toward the southwestern Atlantic ocean, specifically influences climatology of the studied watershed along the October to March time frame as it does for the whole southeast region in Brazil. The area is also influenced by large-scale phenomena related to sea surface temperature (SST) variability, such as El Niño Southern Oscillation (ENSO), hot phase (El Niño—EN), and cold phase (La Niña—LN), jointly with the Atlantic Multidecadal Oscillation (AMO). Front systems, originated in the South Pole, also contribute to enhance precipitation amounts over the watershed.

Both ENSO and SACZ are said to be responsible for inter-annual precipitation variability and yearly amount climatology in the southeastern region in Brazil, higher than 1500 mm over most of the study area.

A discrete wavelet transform is applied to monthly precipitation time series for periodicity analysis. Satellite-based precipitation estimates are compared to observational monthly precipitation for spatial representativeness and homogeneity evaluation.

Similarity among temporal distribution of point observed precipitation amounts and satellite estimates is verified, leading to the conclusion that, at least in the studied region, remote sensing estimates of precipitation, especially for longer durations, may be adequate to reproduce the observed precipitation pattern and its impacts with respect to vegetation diversity and density.

The identification of hydrological patterns based on vegetation indices under non-stationary conditions is still an ongoing scientific question to be better addressed at the watershed scale. In this work, vegetation cover dynamics is evaluated using the rescaled range analysis (R/S) for Hurst coefficient technique applied to normalized difference vegetation index (NDVI) bimonthly time series adopted as an indicator of vegetation cover state.

The NDVI values follow the variability of precipitation and show persistency throughout most of the watershed, as the Hurst exponent values indicate. Non-persistent NDVI is located on lower and flatter relief with different hydrogeological formation and vegetation cover, notably grassland of a less dense type, and higher H values occur along forested areas limited by the annual 1400 mm contour on the isohyet map.

The Morlet continuous wavelet transform with spectrum analysis shows a significant 12-month cycle of the annual rainy season and the 3–4 month cycle possibly related to rain distribution along trimesters. If the time series were longer, other cycles of longer time periods would possibly be revealed, and the association with ENSO or with SACZ would have emerged with more evidence.

The possibility of explaining the high annual totals in the region, as a result of the performance of the various mechanisms of precipitation formation, would increase the predictability of rainfall in the basin. The episodes of EN seem to favor the occurrence of SACZ in the southeast with a marked oceanic connection. The absence of these mechanisms, a sequence of neutral ENSO years or the absence of SACZ in the summer, may possibly anticipate prolonged droughts and water crises.

An example could be the development of the EN episode (2015–2016) that was reported as confirmed around July and August 2015. Its first impacts generated consequences as heavy rains over the southeastern region took place in November, although heavy rains had been occurring since July over the southern region.

In the EN cycle, which largely coincides with the hydrological year of the study region, it is also possible to predict that, in the following year, the months of May and June, normally drier, will be rainy, as happened in May and June 2016.

The extreme events (intense rains and prolonged droughts) that occur in the representative basin of Piabanha are largely induced by the same phenomena and meteorological systems that operate in other basins in the southeastern region. Some of these basins have great economic importance, as the one formed by the Atibaia and Capivari rivers, the main source of water supply for the metropolitan region of São Paulo and Campinas, and the Paraíba do Sul river basin, where there is an expressive industrial park, and that is the main source of water supply for the metropolitan region of Rio de Janeiro. This fact reinforces the relevance of the experimental basins of the EIBEX project, whose data and research can support the decision-making processes regarding the proper management of water resources.

The EIBEX project is thus an important laboratory for the southeast region in the investigation and solution or mitigation of its biggest problems, natural disasters, disorderly occupation, and pollution by domestic sewage. Despite real urban problems, the Piabanha basin still retains significant secondary vegetation cover from the Atlantic Forest biome, present in more than 50% of its area.

A final word is devoted to emphasize the necessity of developing further research work to explore the hydrological patterns of floods and droughts based on the hydrometeorological connections between the Amazonia and the southeast of Brazil, as previously commented in the introductory section of this chapter. In particular, more investigations are needed with respect to the SACZ, as it is in fact an important climatologic feature of the austral summer in South America.

The topics discussed above attest the importance of evaluating water resources management and pollution treatment with a full comprehensive approach that allows to more accurately identify hydrological and meteorological patterns as driving forces toward developing improved water and energy balances. Therefore, it is expected that such framework provides adequate assessment and support for the development of policies focused on water security and on wise sustainable environmental conditions for any region worldwide.

3 Case Study 2: Sustainable Practices in the Serra Azul Catchment—Minas Gerais

In this section, the implementation of a program for the provision of ecosystem services is discussed in the sense of conceiving land use and soil occupation scenarios and introducing conservation and environmental restoration practices in a catchment which is an important source of drinking water for a metropolitan region located in the state of Minas Gerais, Brazil. The central concerns of the ecosystem services are related to water supply (regulation of water flow and water quality) and erosion control. In the first scenario analyzed, the protection and recovery of areas that are already legally protected in Brazil but often suffer anthropogenic intervention were sought; in the later scenarios, new conservation strategies are incorporated, considering the environmental fragility of the basin and the land use characteristics. The costs associated with the proposed techniques were then estimated. The evaluation of costs assists in the decision-making process regarding possible alternatives for the provision of ecosystem services, considering the costs associated with each one and providing subsidies for the use of preventive strategies, such as, for example, the establishment of a compensation scheme for ecosystem services seeking environmental conservation. It is an ongoing study in which part of the research objectives have not yet been fully addressed.

3.1 Area of Study

The Serra Azul Reservoir, located approximately 55 km from Belo Horizonte in the Brazilian state of Minas Gerais, is one of the main sources of drinking water for the Belo Horizonte Metropolitan Region (BHMR), with capacity to supply about 270,000 dwellings.

Serra Azul is part of the Paraopeba drinking water system, an integrated system that encompasses the water storage, withdrawal, treatment, and distribution. In addition to Serra Azul, the system consists of two more reservoirs: Rio Manso and Vargem das Flores, and a direct water withdrawal from the Paraopeba River. The four systems are integrated with the Velhas drinking water system, with a total nominal capacity of about 20 m^3/s to supply a population of 5.8 million inhabitants.

The Serra Azul catchment has an area of 263 km^2, of which approximately 30 km^2 constitutes a strict protection area in the vicinity of the reservoir, owned by Copasa, the water utility that operates the system. The catchment covers territories of four municipalities with approximately 87% of its area located in the municipalities of Mateus Leme and Igarapé.

With a traditionally agricultural and mineral (iron ore) vocation, the catchment environmental issues have always been strongly linked to these two sectors. Recently, a third aspect, urbanization, has influenced the local territorial dynamics, due to urban expansion of the cities located in or near the catchment, and the increase in the number of new gated neighborhoods.

Mining, mainly of iron ore, is present along the border of the basin to the south. The mining sector, with activities normally classified as presenting high polluting potential and formed by large companies, is subject to a more rigorous control and inspection of its activities by public environmental agencies. Mining companies usually have a corporate sector of institutional relations that defines the intervention strategies in the territory and relationships with surrounding communities, with incorporation of the demands of the environmental licensing process. A particular context for the mining sector is the environmental and social risk associated with tailings dam breaks. However, in the case of the Serra Azul, it should be noted that tailings dams are not implanted in the area.

The agricultural sector is extremely important in the Serra Azul catchment. It has relevance as a source of income for rural producers and for the supply to BHMR, mainly of horticulture products. Agricultural activity is carried out predominantly in small family farms of 5 to 8 hectares. In cultivation, traditional management techniques are used, with intensive use of agrochemicals. According to Copasa technicians, the water in the catchment is considered of good quality and conventional water treatment guarantees the water potability. However, surface runoff in agricultural areas where pesticides are used introduces the risk of micropollutant contamination. In Brazil, monitoring of micropollutants in sewers and waters is still incipient and biological sewage treatment techniques are not efficient to remove these substances. In addition, legislation itself lists only a few parameters denominated "micropollutants" or "emerging contaminants." It is possible that water considered potable may present contamination by substances for which there is no legislation although potentially harmful to the ecosystem and human health (Soares and Leão 2015).

Production on agro-ecological bases is practically non-existent in the catchment. According to interviews with agronomists working in the area, farmers are resistant to the use of "alternative" technologies, not part of the local "culture." Recently, many of the farmers changed irrigation systems, migrating from the micro sprinklers to the drip irrigation techniques, which require less water consumption. This change occurred due to economic and financial reasons in the face of recent periods (2014–2015) of water scarcity. The change was perceived as the only possible alternative of economic viability for farming. For this same reason, farmers have migrated to new crops, such as cassava and gannica, which are less water-intensive. Rural producers are presently the main target population of an existing program in Serra Azul focused on monetary payment for ecosystem services, namely the Igarapés Guardian Project ("Guardião dos Igarapés"), developed by the municipality of Igarapé with support from the National Water Agency. Copasa also develops environmental recovery and education actions carried out in collaboration with municipalities and partner entities.

Urban expansion areas of Igarapé, farms, and condominiums for residential use complement the mosaic of anthropic uses in the catchment, which still has an expressive percentage of areas covered with native vegetation (66% of the area). Even outside the protected area of Copasa, there are large portions with remnants of native forest. The presence of this still expressive percentage of vegetated area seems to be related to the low attractiveness of the catchment to the real state market. The major expansion areas of the BHMR are closer to Belo Horizonte, going through an expansion process of the urban fabric of the capital. But in addition to the farms and leisure sites, the region has been the target of new real state parceling, many equipped with irregular groundwater withdrawal systems.

Given this scenario, negotiated territorial approaches for protection of drinking water (preventive solution) are alternatives to the increasing complexity of water treatment processes to meet health standards required by legislation (curative solution), or to resort to other solutions, such as the increase of existing funding, searching for new sources or interconnection between them (palliative solution).

Water scarcity leads to the need to think of ways to conserve water. It is assumed that there is potential for alteration on land use in the Serra Azul catchment, with a risk of deforestation of remaining native woods, which reinforces the need for preventive measures to avoid water quantity and water quality problems to the reservoir or even the shortening of its useful life.

3.2 *Scenarios for Cost Analysis and Evaluation*

Simulations were carried out using four scenarios of land use in the catchment. The scenarios analyzed are based on current land use (considering the year 2016) and progress toward more conservationist scenarios. In this stage of the work, the alternatives that would occur in the case that legislation was effectively fulfilled or the appropriate technical solutions from the physical-biological point of view were used, even effectively incorporating the participation of the social segments involved, which would allow co-responsibility of social actors in relation to activities and projects. The institutional framework for the management of services, cultural and social aspects will be addressed in the second stage of the research.

To identify the best technique for recovery of the area and the activities involved, a structured diagnosis of the catchment is required, consisting of an analysis of its environmental resources and their interactions. Aspects such as the land use and soil types and conservation status, presence of native remaining forest, topography, water regime, history of land use, and socioeconomic profile are analyzed. In this work, considering the adopted scale for investigation, macrospacial indicators and average parameters are used to estimate recovery costs. The objective here is not to undertake an analysis at the executive project level, but to obtain information for management and planning of environmental actions in the catchment.

- **Scenario C 1.0—Current**

A land use map (Fig. 24) was obtained by Matos et al. (2017) from images of the Sentinel 2 satellite acquired from year 2016. The area around the reservoir, which delimitates its direct runoff contributing area, is classified as a Special Protection Area (SPA) and owned by Copasa, the water utility that operates in the region of study. SPA is a strictly protected area used only for the purpose of preservation and environmental protection.

- **Scenario C 1.1—Legal Green**

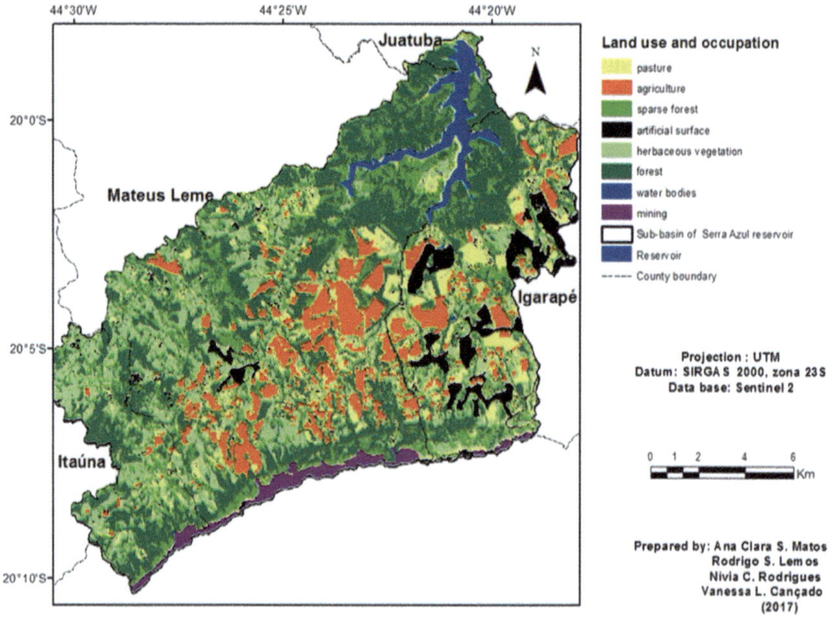

Fig. 24 Land use and vegetation cover in the Serra Azul reservoir basin (Matos et al. 2017)

This scenario reproduces the legal considerations of the Brazilian Forest Code in relation to Permanent Preservation Areas (PPAs) respecting riparian areas, hilltops, hillsides with accentuated slopes and areas surrounding water springs. They have environmental functions of preserving water resources, landscape, geological stability and biodiversity, facilitating the genetic flow of fauna and flora, and protecting the soil and ensuring well-being of human populations.[5] The PPAs account for approximately 12.21% of the total catchment area, where 2.08% present anthropic use, i.e., they are in an irregular situation. The impacted PPAs were selected as areas of the highest priority for recovery or conservation, in view of their environmental functions and current legislation.[6]

The techniques of forest restoration vary from those with none or low direct interventions to those having a high degree of intervention. Nucleation approaches were here selected including the creation of nuclei at strategic points from which the restoration radiates to occupy areas without vegetation. In these nuclei, planting a mixture of native tree species is carried out using the characteristics of the flora in the studied catchment as a reference, which acts as catalysts of ecological succession to promote return to the Seasonal Forest (Parrotta et al. 1997; Moraes et al. 2013).

For estimation of the PPA recovery cost, the actions considered were: the plantation, replanting/maintenance in the 1st year, replanting/maintenance in the 2nd year and fencing costs to which management costs were added. The performance of each of these activities is conditioned to the type of land use and the PPA type (riparian areas and springs). The cost structure, including the monetary values for each activity and input, is obtained from budgets of the Degraded Areas Recovery Plans (Sete Soluções e Tecnologia Ambiental 2017). Not included in the analysis were PPAs located in mining areas, in urbanized areas or occupied by rural infrastructure, and those located in the area under the responsibility of Copasa, due to the legal, technical, and environmental specificities involved.

- **Scenario C 1.2—Legal Green Plus**

This scenario is the conjunction of Scenario 1.1 with forest restoration in areas highly susceptible to erosion (approximately 8.09% of the catchment area) and the preservation zones according to the land use regulation proposed in the BHMR Development Plan (PDDI, acronym in Portuguese). One of the main tools of the

---

[5] In Brazil, PPAs were established as defined in the Law on Native Vegetation Protection ("New Forest Code"), Law no. 12.651/2012 (Brazil 2012). The PPAs must be occupied by native vegetation maintained by the owner of the area, possessor or occupant. Interventions or suppression of native vegetation in PPAs, provided by law, occur in restricted situations, in the cases of public utility, social interest or low environmental impact. Despite legislation, alternative uses in PPAs are frequent. Today, a large part of the environmental programs for the recovery of catchments and river basins in Brazil are focused on the recovery of PPAs.

[6] "Legal Reserves", although they are also classified as priority intervention areas, were not considered due to the lack of information about them in the Serra Azul catchment. According to the Brazilian Forest Code, all rural properties must maintain an area with native vegetation cover, as a Legal Reserve, without prejudice to application of the rules on Permanent Preservation Areas. In the state of Minas Gerais, the percentage dedicated to the Legal Reserve is 20% of the total area of the property, and the law includes exceptions, as in the cases of small rural properties.

PDDI land use regulation framework is the land use zoning. Considering the large scale of metropolitan regulation, this tool is called a macrozone. When land use regulation is justified by the metropolitan environmental interest of the area, such as catchments which are strategic water sources for the BHMR, the so-called Protection Macrozone 1 in this legislation is the more restrictive. In the case of Serra Azul, 38.21% of the catchment area is classified as Protection Macrozone 1, restricting the occupation to those that contribute to environmental conservation and recovery. The remaining area is classified as Protection Macrozone 2, allowing agricultural exploitation, and Protection Macrozone 3, accepting urban developments with low impervious surface occupation, limited to 20% of the lot surface.

In addition to the costs analyzed in Scenario 1.1, the costs of forest restoration in the areas mentioned above are included here through the low-cost nucleation techniques.

- **Scenario C 1.3—Soil and water conservation**

This is the combination of Scenario 1.2 with practices that seek to conserve water and soil. No-tillage in the agricultural area and terracing in the pasture area were considered. No-tillage is a planting technique in which the seed is placed in the soil without plowing (without prior aeration or tilling) using special planting machines. The technique is considered economically and environmentally efficient (Jardim 2010). Terracing is one of the most widespread practices among Brazilian farmers for the control of soil erosion due to rainfall-runoff processes, basically consisting of the construction of terraces, structures arranged transversely to the slope of the terrain, aiming to promote infiltration and to reduce surface runoff velocity and its potential to causing erosion (Vital and Resck 2002).

The cost parameters for terrace implementation were obtained with reference to the basic terracing project of the Igarapés Guardian Project, currently under development in the Serra Azul catchment.

This is still a preliminary estimate for scenario C1.3, since it does not incorporate the maintenance costs of the terraces, nor does it estimate the cost of no-tillage compared to traditional planting.

An estimation of the benefits to the hydrological cycle and for erosion control and sediment yield resulting from adoption of the described scenarios was performed by hydrological modeling using the SWAT model (Arnold et al. 2012), and in the case of sediment production by means of the RUSLE equation—Revised Universal Soil Loss Equation (Renard et al. 1997; Fu et al. 2006). Results of these simulations, fundamental to justify the potential adoption of the scenarios considered herein, can be found in Laender (2018).

3.3 Results

The results regarding the four scenarios are depicted in Fig. 25. The present vegetated area of the catchment (C1.0), considering the dense, sparse, and herbaceous vegetation cover is 66.5% and increases to 68.6% in the case of C1.1. This slight increase of 2.1%, due to the restoration of the PPAs, is a consequence of the already extended vegetation cover in Serra Azul catchment. In C1.2, the vegetated area increases to 72.6%, i.e., an increase of 6.1% in relation to the current scenario. This increase is due, in addition to the restoration of the PPAs, to the restoration of the most environmentally vulnerable areas of the catchment.

Although the vegetated area in C1.2 and C1.3 are the same, scenario C1.3 is considered to have the greatest conservation potential, since, in addition to the extension on the vegetation cover, it incorporates soil and water conservation measures in farming practices.

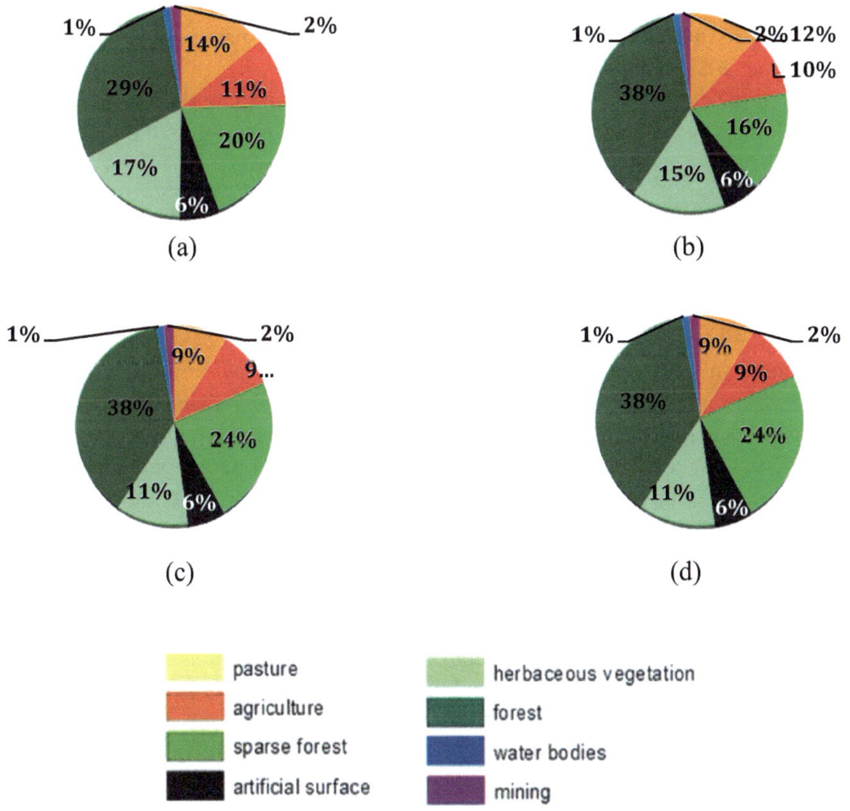

Fig. 25 Distribution of the typologies of use and vegetative cover for the scenarios C1.0 (**a**), C1.1 (**b**), C1.2 (**c**) and C1.3 (**d**)

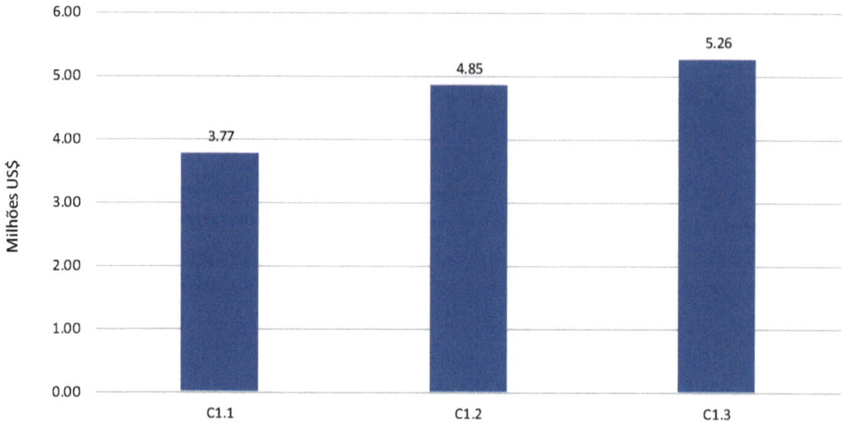

Fig. 26 Total costs of the proposed measures, per scenario, in US$

Figure 26 shows the investment values for each scenario, based on cost assessment and taking 2018 as time reference.

Given the characteristics of the territory, one question that arises is the involvement of the various actors, in view of the possible tools to promote the provision of ecosystem services, as discussed in the following paragraphs.

- Collectivity: The hypothesis here is that the collectivity (water users) supplied by the Serra Azul drinking water system will fund the investments in the catchment protection through the water bill. Values of the proposed measures imply annual cost per household supplied by the Serra Azul system of US$ 1.87 for C1.1, US$ 2.40 for C1.2, and US$ 2.61 for C1.3. The financial calculations to arrive at these values were conducted according to the following premises: investment for environmental recovery would be financed over 12 years; the Serra Azul system serves approximately 266 thousand households; the interest rate adopted is Copasa's capital remuneration (7.94% per year) and the amounts are present values, i.e., they have been updated to US$ of 2018. In this case, the collectivity, as a user of the water produced in the basin, would be paying for the provision of ecosystem services that would assist in its protection. It can be argued that this form of financing is in contradiction with the "polluter pays" principle of economic rationality and that the consumer should be entitled to clean water since he/she is not the direct agent of catchment degradation. There is a sense of public good related to water, and that other agents should pay for these costs, be it the public power, the rural owners or Copasa, each for different reasons. But this solution can be plausible if the actions are taken transparently, with a well-informed consumer who is willing to pay that amount in exchange of water security. A fundamental aspect is to enable the monitoring of conservation measures and environmental recovery by the community. Ideally, communities would be integrated in the process from the outset, including participation in the selection of techniques used in the catchment. Another point is that, in principle, there is no justification for such investments being made in this catchment and not in another, with more critical environmental quality. It is an alternative that should be considered through an integrated analysis of prioritization of investments in catchments that are sources of drinking water, and the family budget may be a possible source of financing, among others. In any case, the small share of the cost transferred to the family budget would assist in the implementation of a program of shared benefits between Copasa and the population served by the system. Two caveats must be made: the opportunity cost incurred by rural producers was not analyzed, which must be relatively high, as well as the transactional costs involved in implementing and maintaining the payment for ecosystem services framework.

- Water utility: The water utility is a relevant actor in the catchment, since potable water is the most valuable asset in the area. Better quality of raw water means lower costs for treatment. A trade or negotiation between Copasa and the actors that interfere with water production and quality, such as rural producers, appears to be realistic. Copasa conducts environmental education programs and recovery of springs, soils, and riparian forests in Serra Azul and is responsible for the protection of a huge forested area surrounding the reservoir. It leads a program for the distribution of native tree seedlings to farmers, with guidance on the correct way of planting and the necessary care for the development of the species. Farmers are also made aware of the importance and necessity of native vegetation for spring and riparian area preservation. One of the company's principles is the non-use of monetary payment for the provision of ecosystem services, mainly seeking engagement of the local population in environmental protection and recovery measures.
- *Municipal governments,* via *public budgets*: The municipal government of Igarapé conducts the Igarapés Guardian Project, which has several partners, including Copasa. It includes 22 rural properties, located in a priority sub-catchment for recovery. For each property, a specific environmental re-adaptation project is elaborated, for which there are proposals for forest restoration, soil conservation (adaptation of roads and mines), and sanitation (water tanks and domestic sewage). The actions are carried out by the project team, without burden to the owner, with resources from the National Water Agency (federal agency). The program also includes cash payment to rural landowners (via municipal budget) with values that vary according to the size of the property, biodiversity (forest cover area), and good practices (sustainable agriculture and environmental sanitation). Because owners and rural workers are made aware of good practices in rural areas, including the non-use of fire for the clearing of arable land, they are invited to be volunteer brigaders, reinforcing the idea that they are true guardians of the igarapés. The program encompasses integrated measures, differentiation of "environmental benefits," and is concerned with the engagement of social groups. It has the disadvantage of high transactional costs, since a project is done with each owner, suitable for small critical intervention areas.
- *Rural landowners*: in Scenario C1.1, the conservation of native vegetation in PPAs is a legal requirement, therefore public authorities can require the restoration of

these areas by rural landowners. However, in Brazil, anthropic interventions in PPAs are usual, due to lack of knowledge on legislation by property owners or because it is something "normal" within the collective consciousness, due to the historical absence of an authority capable of constraining intervention in these spaces. In addition, because it is a catchment made up of small farms with relevant economic output, the use of part of the property for protection purposes may impact farmers' income. There is also concern about the availability of water, since agricultural irrigation is practiced with high water consumption, even by dripping. Illegal drilling of wells and non-registered water withdrawal make it difficult to estimate the volumes collected. Demanding that property, owners assume the cost of readjusting their properties according to the legal norm tending to have a high social and political cost, with potential socio-environmental conflicts. According to interviews conducted with rural extension agents working in the catchment, rural producers tend to favor the short term from an economic-financial perspective. The hypothesis of this work is that they would more willingly accept a better system to stimulate the provision of ecosystem services, either via cash payment or indirectly, through in-kind compensations.

- *Mining companies*: in Brazil, the path to obtaining licenses for the implementation and operation of mining enterprises is long and bureaucratic, with the need for relatively detailed studies of environmental impact and requirements for programs and actions to minimize, mitigate, or compensate for the predicted impacts. Mining companies have great economic power, act in a restricted area of the catchment and are subject to the legal obligations inherent to the process of obtaining licenses (control of atmospheric emissions and generation of effluents, monitoring, among other issues). With a specific line of action in the territory, the mining sector is not always a participating actor in the integrated environmental management decisions in the catchment.
- *Other actors*: Federal, State, Non-Governmental Organizations, Cooperatives, among others. In the basin, there is a cooperative of rural producers, with limited actuation, but important for the articulation with producers. At the federal level, the National Water Agency has designed and conducts the Water Producer Program, which supports projects seeking to improve the water resources of a river basin in a joint action with public and private institutions operating in the region, organized in a way that each can contribute with actions of its routines without having to expend extraordinary efforts or resources. The idea is based on the principle that there is a group of people interested in water protection and willing to contribute, while, on the other hand, there are rural producers who can integrate and enable the conservation of water resources through proper management of their properties, transforming them into providers of environmental services (ANA, 2018).

4 Concluding Remarks

This chapter investigated Brazilian hydrological experiences from Piabanha river watershed experiments in Rio de Janeiro and from Serra Azul reservoir catchment in Minas Gerais toward new policies for planning megacities.

In the first case study, as presented in Sect. 3, a representative basin with its experimental watersheds is used as a laboratory for testing methodological approaches and innovative initiatives proposed to analyze and mitigate megacities issues located in the Brazilian southeast region. The basin has intense urbanized areas, but preserves part of the original vegetation cover of the Atlantic rainforest biome. The humid climate under the activity of cold fronts, of the South Atlantic Convergence Zone (SACZ) and of other meteorological systems and climate drivers, where the annual precipitation mean exceeds 1500 mm, 50% of it concentrated along the summer months, makes the region prone to natural disasters aggravated by urbanization. The main purpose of this research was to identify patterns at different levels and dimensions, notably multi-scales (urban and watershed) and multi-utilities, which could help in the diagnosis and in systematizing proposals for mitigating water-related issues, in special, floods, flashfloods, and water supply crisis for risk reduction. Available satellite products and databases, such as vegetation spectral indices, precipitation, digital terrain models, spatial distribution of population and occupation aspects are analyzed using data science tools. The findings of the research are verified for the area of two megacities in Brazil, namely São Paulo and Rio de Janeiro metropolitan regions, southeastern Brazil, showing to be reproducible for those environments.

In the second case study, depicted in Sect. 3, when estimating costs in different catchment conservation and recovery scenarios, we obtained indicators that can be analyzed in conjunction with those of other solutions, preventive, palliative, and curative. To continue the study, calculation of the opportunity cost is an important step, since forestry restoration in areas of economic use or conservation of native vegetation in areas that could potentially have other uses implies a loss of income (real or potential) to the owner, for the benefit of the community.

Serra Azul is a catchment located in a rural and peri-urban area, providing water mainly for metropolitan urban area residents. This is an issue that involves solutions of compromise between elements of demographic pressure, land use, agricultural and mining economic interests, water utility, and urban collectivity.

Local environmental culture does not yet consider broader negotiations, with rural landowners as "guardians" of biodiversity. But there is potential in new forms of territorial intervention, where environmental and economic objectives are not intrinsically conflicting, but, on the contrary, can guarantee solutions that allow for the sustainability of agricultural activity in the long term, or even an incremental income. Agroecological practices, for example, can increase the value of agricultural products when marketed as "organic," increasing the income of agricultural producers. Recovered areas can supply the demand for rural tourism, an activity that can be a factor of income generation with reduced environmental impact in these areas. Agroforestry systems have also been proposed by several authors as one of the best alternatives for land use in the search for the reconciliation of economic and ecological benefits, particularly in regions that have already been deforested, degraded, unproductive, or permanently occupied with agriculture or pasture (McCaffery et al. 2002 cited in Neto 2008). This combination of livelihood improvement together with the provision of a critical resource, such as water, seems to be an important driver of the success of PES schemes (Grima et al. 2016).

To summarize, we hope this chapter brings new insights to the development of innovative approaches to deal with hydrological change and water security based on basin in situ instrumentation and remotely sensed data acquisition and analyses jointly with hydrological modeling.

Acknowledgements The authors would like to thank the Civil Engineering Program of the Alberto Luiz Coimbra Institute for Graduate Studies and Research in Engineering (COPPE)—Federal University of Rio de Janeiro (UFRJ) through the support of the Laboratory of Water Resources and Environmental Issues (LABH2O) and the Graduate Program on Sanitation, Environment and Water Resources—Federal University of Minas Gerais (UFMG) through the Department of Hydraulic Engineering and Water Resources with respect to data and infrastructure provided by these institutions for this research. The authors would like to express their sincere appreciation to the financial support for research projects, which came through Brazilian agencies, notably CAPES (Coordenação de Aperfeiçoamento de Pessoal de Nível Superior), CNPq (Conselho Nacional de Desenvolvimento Científico e Tecnológico), FAPERJ (Fundação de Amparo à Pesquisa do Estado do Rio de Janeiro, FAPEMIG (Fundação de Amparo à Pesquisa do Estado de Minas Gerais) and FINEP (Financiadora de Estudos de Projetos). CAPES also funds UFMG and UFRJ participation in the BRICS-NU framework. The first and second authors are supported by CNPq under the grant program CNPq-PQ. The authors would like also to acknowledge ANA (Brazilian National Water Agency), CPRM (Brazilian Research Company of Mineral Resources)—SGB (Brazilian Geological Survey), COPASA (Minas Gerais State Water Utility Company), INPE (Brazilian National Spatial Research Institute), EMBRAPA (Brazilian Agricultural Research Corporation), CEPEL (Brazilian Electrical Engineering Power Research Center), ONS (Brazilian National Electric System Operator) and INMET (Brazilian National Institute of Meteorology) for the continuous support of the research in hydrology in Brazil.

References

Aitken AP (1973) Assessing systematic errors in rainfall-runoff models. J Hydrol 20:121–136

Allen RG, Pereira, LS, Raes D, Smith M (1998) Crop evapotranspiration - guidelines for computing crop water requirements - FAO Irrigation and Drainage Paper N° 56. FAO, Rome, 300 p

Araujo AAM (2006) Um novo esquema de parametrização hidrológica da superfície terrestre com redistribuição lateral da água no solo (A new hydrologic land surface parametrization scheme with lateral soil water redistribution). D.Sc. Thesis, Civil Engineering Program, COPPE/UFRJ, Rio de Janeiro, RJ, Brasil, 236p

Araujo LMN (2016) Identificação de padrões hidrológicos de precipitação e de umidade do solo na bacia hidrográfica do rio Piabanha/RJ (Identification of precipitation and soil moisture hydrological patterns at Piabanha/RJ river basin). D.Sc. Thesis, Civil Engineering Program, COPPE/UFRJ, Rio de Janeiro, RJ, Brasil, 236p

Arnold JG, Kiniri JR, Srinivasan R, Williams JR, Haney EB, Neitsch SL (2012) Soil water assessment tool: input, output documentation, version 2012, Texas Water Research Institute TR 439 - Tamu

Beven KJ (1989) Changing ideas in hydrology: the case of physically-based models. J Hydrol 105:157–172
Beven KJ, Binley A (1992) The future of distributed models: model calibration and uncertainty prediction. Hydol Process 6:279–298
Beven KJ, Freer F (2001) Equifinatility, data assimilation, and uncertainty estimation in mechanistic modelling of complex environmental systems using the GLUE methodology. J Hydrol 249:11–29
Beven KJ, Lamb R, Quinn P, Romanowicz R, Freer F (1995) TOPMODEL. In: Singh VJ (ed) Computer models of watershed hydrology. Water Resources Publications, Colorado, USA, pp 627–668
Blöschl G et al (2019) Twenty-three unsolved problems in hydrology (UPH) a community perspective. Hydrol Sci J 64(10):1141–1158. https://doi.org/10.1080/02626667.2019.1620507
Bobbio N (2011) Da estrutura à função: novos estudos da teoria do direito, São Paulo, Editora Manole Ltda
Bracer C, Waage S, Inbar M (2007) Getting started: an introductory primer to assessing & developing payments for ecosystem service deals. Washington, DC, Katoomba Group. Available at: http://www.katoombagroup.org
Braga BPF, Flecha, R, Pena DS, Kelman J (2008) Federal pact and water management, Estudos Avançados 22(63):17–42
Castro A, Verburg P, Martín-López B, Garcia-LLorented M, Cabello J, Vaughn C, López A (2014) Ecosystem service trade-offs from supply to social demand: a landscape-scale spatial analysis. Landsc Urban Plan 132:102–110
Chamoli A, Bansal AR, Dimri VP (2007) Wavelet and rescaled range approach for the Hurst coefficient for short and long time series. Comput Geosci 33:83–93
CPRM (2002) Regionalização de Vazões da Sub-Bacia 58 – rio Paraíba do Sul – Convênio ANEEL/CPRM. Rio de Janeiro
CPRM (2011) Isoietas Médias Anuais de 1977 a 2006 na escala 1:5.000.000. In: Atlas Pluviométrico do Brasil. Maceió. CD e http://www.cprm.gov.br/publique/Hidrologia/Mapas-e-Publicacoes/Atlas-Pluviometrico-do-Brasil-1351.html. Accessed in: 2 April 2015
CPRM (2014) Mapa Hidrogeológico do Brasil ao Milionésimo: Sistema de Informação Geográfica – SIG Escala 1:1.000.000. Disponível em: http://www.cprm.gov.br/publique/Hidrologia/Mapas-e-Publicacoes/Mapa-Hidrogeológico-do-Brasil-ao-Milionésimo-756.html. Accessed in: 2 Feb 2016
CPRM (2015) Rede Integrada Monitoramento das Águas Subterrâneas – RIMAS. Disponível em: http://rimasweb.cprm.gov.br/layout/. Accessed in: 25 Oct 2016
Dafne Chan P, Costa FS, Damázio JM, Rotunno Filho OC (2019) Stationary of extreme flow time series for flood control operation in hydropower reservoirs, Cadernos do IME – Série Estatística, Universidade do Estado do Rio de Janeiro – UERJ – ISSN impresso 1413–9022/ISSN on-line 2317–4355, v. 46, pp 16–29. https://doi.org/10.12957/cadest.2019.50174
Daubechies I (1992) Ten lectures on wavelets. SIAM, Philadelphia, PA, pp 278–285
Di Baldassarre G, Viglione A, Carr G, Kuil L, Salinas JL, Blöschl G (2013) Socio-hydrology: conceptualising human-flood interactions. Hydrol Earth Syst Sci 19:3295–3303
Di Bello RC (2005) Análise do comportamento da umidade do solo no modelo chuva-vazão SMAP II – versão com suavização hiperbólica – Estudo de caso: região de Barreiras na bacia do rio Grande-BA (Analysis of the soil moisture behavior in the rainfall-runoff SMAPII model – hyperbolic smoothing version – Case study: Barreiras region in Rio Grande watershed – Bahia, Brazil). M.Sc. Thesis, Civil Engineering Program, COPPE/UFRJ, Rio de Janeiro, RJ, Brasil, 225 p
Dooge JCI (1986) Looking for hydrologic laws. Water Resour Res 22(9):46S-58S
Dooge JCI (1997) Scale problems in hydrology. In: Reflections on hydrology. American Geophysical Union, 314 pp
Duan Q, Sorooshian S, Gupta V (1992) Effective and efficient global optimization for conceptual rainfall-runoff models. Water Resour Res 28:1015–1031

Ehsani N, Vörösmarty CJ, Fekete BM, Stakhiv EZ (2017) Reservoir operations under climate change: storage capacity options to mitigate risk. J Hydrol 555:435–446
Enriquez RP (2014) Programa en matlab para el calculo del exponente de Hurst y dimension fractal de series de tempo, Centro de Geociencias, UNAM. Mexico. Available at: http://www.geociencias.unam.mx/~bole/eboletin/progHurstyD1108.pd. Accessed in 1 April 2014
EPA - UNITED STATES ENVIRONMETAL PROTECTION AGENCY (2009) Valuing the protection of ecological systems and services: a report of the EPA science advisory board. Washington, DC, Science Advisory Board, Office of the Administrator
Forest Trends (2015) Incentivos Econômicos para Serviços Ecossistêmicos no Brasil, Rio de Janeiro
Fu G, Chen SL, Mccool DK (2006) Modeling the impacts of no-till practice on soil erosion and sediment yield with RUSLE, SEDD, and ArcView GIS. Soil Tillage Res 85(2):38–49
Gesisky J (2017) Código Florestal Brasileiro: haverá futuro? WWF, Brasil, Brasília (DF)
Getirana ACV, Bonnet MP, Calmant S, Roux E, Rotunno Filho OC, Mansur WJ (2009) Hydrological monitoring of poorly gauged basins based on rainfall-runoff modeling and spatial altimetry. J Hydrol 379:205–219
Getirana ACV, Bonnet MP, Rotunno Filho OC, Collischonn W, Guyot J-L, Mansur WJ (2010) Hydrological modeling and water balance of the Negro river basin: evaluation with observed data and spatial altimetry. Hydrol Process 24:3219–3236
Getirana ACV (2009) Contribuições da altimetria espacial à modelagem hidrológica de grandes bacias na Amazônia (Spatial altimetry contributions to the hydrological modeling of large basins in Amazon). Ph.D. Thesis, Civil Engineering Department, COPPE/UFRJ, Rio de Janeiro, RJ, Brasil, 273p
Gonçalves RC (2008) Modelagem hidrológica do tipo chuva-vazão via SMAP e TOPMODEL – Estudo de caso: bacia do rio Piabanha (Rainfall-runoff hydrological modeling using SMAP and TOPMODEL – Case study: Piabanha river watershed/RJ). M.Sc. Thesis, Civil Engineering Program, COPPE/UFRJ, Rio de Janeiro, RJ, Brasil, 239 p
Grima N, Singh S, Smetschka B, Ringhofer L (2016) Payment for ecosystem services (PES) in Latin America: analyzing the performance of 40 case studies. Ecosyst Serv 17:24–32
Grinsted A, Moore JC, Jevrejeva S (2004) Application of the cross wavelet transform and wavelet coherence to geophysical time series. Nonlinear Process Geophys 11:561–566
Groot R (2006) Function-analysis and valuation as a tool to assess land use conflicts in planning for sustainable, multi-functional landscapes. Landsc Urban Plan 75:175–186
Groot R, Wilson M, Boumans R (2002) A typology for the classification, description and valuation of ecosystem functions, goods and services, ecological economics, special issue: the dynamics and value of ecosystem services: integrating economic and ecological. Perspectives 41:393–408
Guedes F, Seehusen S (2011) Pagamentos por Serviços Ambientais na Mata Atlântica: lições aprendidas e desafios, MMA, Brasília
Gupta VK, Sorooshian S (1985) The automatic calibration of conceptual catchment models using derivative-based optimization algorithms. Water Resour Res 21:473–485
Haar A (1910) Zur theorie der orthogonalen funktionensysteme. Mathematische Annalen 69(3):331–371. https://doi.org/10.1007/BF01456326
Hercowitz M (2011) Teste de viabilidade do uso de leilão reverso como mecanismo para pagamentos por serviços ambientais, Projeto de Recuperação de Matas Ciliares, Secretaria de Estado do Meio Ambiente do Estado de São Paulo, Produto técnico número 04
Hilbert D (1902) Mathematical problems. Bull Am Math Soc 8(10):437–479. https://doi.org/10.1090/S0002-9904-1902-00923-3
Hurst HE (1951) Long term storage capacity of reservoirs. Trans Am Soc Civil Eng 116:770–880
Hurst HE (1956) Methods of using long-term storage in reservoirs. Part 1 Proc Inst Civ Eng Part I 519p
IPCC (2007) Climate change 2007:the physical science basis. In: Solomon S, Qin D, Manning M, Chen Z (eds) Contribution of working group I to the fourth assessment report of the intergovernmental panel on climate change. Cambridge University Press, Cambridge

IPCC (2013) Climate change 2013:the physical science basis. In: Stocker TF, Qin D, Plattner GK, Tignor M (eds) Contribution of working group I to the fifth assessment report of the intergovernmental panel on climate change. Cambridge University Press, Cambridge
IPCC (2016) IPCC expert meeting on scenarios. IPCC Working Group III Technical Support Unit, Potsdam Institute for Climate Impact Research, Germany
Irigaray L (1993) An Ethics of sexual difference, translated by Carolyn Burke and Gillian C. Gill, 217 p, Cornell University Press, New York
Jardim MH (2010) Pagamentos por Serviços Ambientais na Gestão de Recursos Hídricos: O Caso do Município de Extrema –MG. M.Sc. Thesis, Centro de Desenvolvimento Sustentável, Universidade de Brasília, Brasília (DF)
Jensen KH, Mantoglou A (1992) Future of distributed modelling. Hydrol Process 6:255–264
Johnston PR, Pilgrim DH (1976) Parameter optimization for watershed models. Water Resour Res 12:477–486
Julián M, Alcaraz R, Rieta JJ (2013) Study on the optimal use of generalized Hurst exponents for noninvasive estimation of atrial fibrillation organization. Comput Cardiol 40:1039–1042
Katsev S, L'Heureux IL (2003) Are Hurst exponents estimated from short or irregular time series meaningful ? Comput Geosci 29:1085–1089
Kavetski D, Kuczera G (2007) Model smoothing strategies to remove microscale discontinuities and spurious secondary optima in objective functions in hydrological calibration. Water Resour Res 43:W03411
Kenter J, O'Brien L, Hockley N, Ravenscroft N et al (2015) What are shared and social values of ecosystems? Ecol Econ 111:86–99
Kitanidas PK, Bras RL (1980) Real-time forecasting with a conceptual hydrologic model, 1, analysis of uncertainty. Water Resourc Res 16:1025–1033
Klemes V (1983) Conceptualization and scale in hydrology. J Hydrol 65:1–23
Klemes V (1986) Dilettantism in hydrology: transition or destiny? Water Resour Res 22(9):177S-188S
Koutsoyiannis D (2003) Climate change, the Hurst phenomenon, and hydrological statistics. Hydrol Sci Journal–des Sciences Hydrologiques 48(1):3–24
Kuczera G (1983) Improved parameter inference in catchment models, 1, evaluating parameter uncertainty. Water Resour Res 19(5):1151–1162
Labat D (2005a) Recent advances in wavelet analysis: part I—a review of concepts. J Hydrol 314:275–288
Labat D (2005b) Recent advances in wavelet analysis: part II—Amazon, Parana, Orinoco and Congo discharges time scale variability. J Hydrol 314:289–311
Laender FR (2018) Modelagem hidrológica com o swat integrando cenários de ocupação do território e técnicas de manejo do solo na bacia de Serra Azul – MG. M.Sc. Thesis; Graduate Program on Sanitation, Envionment and Water Resources, UFMG, Belo Horizonte
Leconte R, Brissette F, Galarneau M, Rousselle J (2004) Mapping near-surface soil moisture with RADARSAT—a synthetic aperture radar data. Water Resour Res 40:1029–1038
Lei no. 12.727 (2012) Alteração da Lei no 12.651, de 25 de maio de 2012, que dispõe sobre a proteção da vegetação nativa. Diário Oficial da República Federativa do Brasil
Libos MIPC (2008) Modelagem hidrológica quali-quantitativa: estudo de caso da bacia hidrográfica do rio Manso – MT (Quality-quantitative hydrological modelling: study case of Manso watershed, MT-Brazil). D.Sc. Thesis, Civil Engineering Program, COPPE/UFRJ, Rio de Janeiro, RJ, Brasil, 367p
Lopes JEG, Braga BPF, Conejo JGL (1982) SMAP, a simplified hydrologic model. In: International symposium on rainfall-runoff modelling. Mississipi State University, Mississipi
Lucena AJ, Rotunno Filho OC, França JRA, Peres LF, Xavier LNR (2013) Urban climate and clues of heat island events in the metropolitan area of Rio de Janeiro. Theoret Appl Climatol 111(3–4):497–511

Luiz Silva W, Xavier LNR, Maceira MEP, Rotunno Filho OC (2018) Climatological and hydrological patterns and verified trends in precipitation and streamflow in the basins of Brazilian hydroelectric plants. Theoret Appl Climatol 1:1–19
Luiz Silva W, Maceira MEP, Rotunno Filho OC (2020) Numerical simulations of precipitation and streamflow in current climate and future projections to drainage areas of Brazilian hydroelectric plants. Climate Res 79:219–241
Mandelbrot BB, Wallis JR (1969) Some long-run properties of geophysical records. Water Resour Res 5:321–340
Mason N, Calow R (2012) Water security: from abstract concept to meaningful metrics: an initial overview of options. Working Paper 357, results of overseas development institute research presented in preliminary form for discussion and critical comment, London
Matos AC, Lemos R, Silva T, Eleutério J, Nascimento N (2017) Evolução do uso e ocupação do solo em mananciais de abastecimento metropolitano na Região Metropolitana de Belo Horizonte, Estado de Minas Gerais, Projeto Modelagem de Mananciais Metropolitanos Estratégicos como Insumo para Gestão de Água e Território em Face à Mudança Climática (report), Programa de Pós-Graduação em Saneamento Meio Ambiente e Recursos Hídricos, Escola de Engenharia, UFMG
May PH (2003) Valoração econômica e cobrança dos serviços ambientais de florestas: identificação, registro, compensação e monitoramento de benefícios sociais. In: Romeiro A (Org.) Avaliação e contabilização de impactos ambientais, Campinas, Editora da Unicamp
Mccafery KA, Fernandes ECM, Rondon MA (2002) Aboveground biomass, carbon and nutrient stocks in three agroforestry systems and secondary forest in the central Amazon, Agroforestry Systems
Medeiros DM, Calmant S, Perosanz F, Xavier LNR, Rotunno Filho OC, Seyler OC, Monteiro A (2016) Comparisons of observed and modeled elastic responses to hydrological loading in the Amazon basin. Geophys Res Lett 9604–9610. https://doi.org/10.1002/2016GL070265
Medeiros DM (2016) Geodésia aplicada ao monitoramento hidrológico da bacia Amazônica (Geodesy applied to hydrologic surveys over Amazon basin). D.Sc. Thesis, Civil Engineering Program, COPPE/UFRJ, Rio de Janeiro, RJ, Brasil, 255p
Millenium Ecosystem Assessment—MA (2005) Ecosystems and human well-being: a framework for assessment. Island Press, Washington, DC, 155 p
Ministry of Regional Development of Brazil—MDR (2018) National water supply and sanitation information system—SNIS. Available at: http://www.snis.gov.br/. Brasília, Brazil
Molinari BS (2015) Modelagem espacial da qualidade de água na bacia do rio Piabanha/RJ (Water quality spatial modeling applied to Piabanha river watershed/RJ). M.Sc. Thesis, Civil Engineering Program, COPPE/UFRJ, Rio de Janeiro, RJ, Brasil, 225 p
Moraes LFD, Assumpção JM, Pereira TS, Luchiari C (2013) Manual técnico para a restauração de áreas degradadas no Estado do Rio de Janeiro, Rio de Janeiro, Jardim Botânico do Rio de Janeiro
Moreira DM, Araujo LMN, Almeida IS (2007) Apoio Cartográfico a Estudos Hidrológicos Utilizando Ferramentas de Geoprocessamento. In: XVII Simpósio Brasileiro de Recursos Hídricos. São Paulo. Anais. 1 CD-ROM
MPOG (2011) Base de informações do Censo Demográfico 2010: Resultados do Universo por setor censitário. IBGE: Rio de Janeiro. Available at: ftp://ftp.ibge.gov.br/Censos/Censo_Demografico_2010/Resultados_do_Universo/Agregados_por_Setores_Censitarios/. Acesso em: 23 April 2013
NASA (2015a) TRMM-based hourly precipitation estimates. Available at: ftp://disc2.nascom.nasa.gov/data/TRMM/Gridded/3B42RT. Access in 24 Aug 2015
NASA (2015b) TRMM-based monthly precipitation estimates. Available at: ftp://disc2.nascom.nasa.gov/data/TRMM/Gridded/3B43_V7/2015/. Access in 1 Oct 2015
NASA (2015c) Frequently asked questions: TRMM spacecraft re-entry. Available at: http://trmm.gsfc.nasa.gov/publications_dir/FAQs_TRMM_Reentry_June2015.html. Access in 1 Oct 2015

NASA (2015d) Transition of 3B42/3B43 research product from monthly to climatological calibration/adjustment. Disponível em: ftp://disc2.nascom.nasa.gov/data/TRMM/Gridded/docs/3B42_3B43_TMPA_restart.pdf. Access in 1 Oct 2015

NASA-GIMMS (2014) NASA global data archive AVHRR Global NDVI (3g) (Normalized difference vegetation index-3rd generation) for 1982–2014

National Water Agency—ANA (2009) Brazilian water resources report—2009: FULL REPORT. Available at: http://www.snirh.gov.br/portal/snirh/centrais-de-conteudos/conjuntura-dos-recursos-hidricos. ANA, Brasília, 204 p

National Water Agency—ANA (2012) Cadastro Nacional de Usuários de Recursos Hídricos HidroWeb, Inventário. http://cnarh.ana.gov.br/. Access in: 1 Oct 2012

National Water Agency—ANA (2013) Brazilian water resources report—2013: FULL REPORT. http://www.snirh.gov.br/portal/snirh/centrais-de-conteudos/conjuntura-dos-recursos-hidricos. ANA, Brasília, 432 p

National Water Agency—ANA (2015) Hidroweb – Sistema de Informações Hidrológicas. https://hidroweb.ana.gov.br/. Accessed in 1 May 2015

National Water Agency—ANA (2017) Brazilian water resources report—2017: FULL REPORT. Available at: conjuntura.ana.gov.br. ANA, Brasília, 169 p

National Water Agency—ANA (2018) Nota informative—Programa Produtor de Água, Brasília. Available at: http://www3.ana.gov.br/portal/ANA/programas-e-projetos/programa-produtor-de-agua

National Water Agency—ANA (2019a) Brazilian national water security plan. Available at: https://pnsh.ana.gov.br/home. Brasília/Brazil, 112 p

National Water Agency—ANA (2019b) Brazilian water uses manual. Available at: http://www.snirh.gov.br/portal/snirh/centrais-de-conteudos/central-de-publicacoes/ana_manual_de_usos_consuntivos_da_agua_no_brasil.pdf. Brasília/Brazil, 76 p

Neto FCVA (2008) Construção dos Mercados de Serviços Ambientais e suas Implicações para o Desenvolvimento Sustentável no Brasil. D.Sc. Thesis, Instituto de Ciências Humanas e Sociais, Curso de Pós-Graduação em Desenvolvimento, Agricultura e Sociedade – CPDA, Universidade Federal Rural do Rio de Janeiro

NOAA-ESRL (2015a) N. Atlantic SST averages, unsmoothed & not detrended. http://www.esrl.noaa.gov/psd/data/timeseries/AMO/. Accessed 27 June 2015

NOAA-ESRL (2015b) HadISST Nino 3.4 SST-NOAA/ESRL PSD. http://www.esrl.noaa.gov/psd/gcos_wgsp/Timeseries/Nino34/. Accessed 1 October 2015

Nusdeo AMO (2013) Pagamento por serviços ambientais. Do debate de política ambiental à implementação jurídica. In: Lavrati P, Tejeiro G (eds), Direito e mudanças climáticas - pagamento por serviços ambientais: fundamentos e principais aspectos jurídicos, 1st. ed., v. 6. Instituto O Direito por um Planeta Verde, São Paulo, pp 8–43. 149 p

O'Donnel T, Canedo PM (1980) The reliability of conceptual basin model calibration. In: Proceedings of the Oxford symposium on hydrological forecasting. IAHS, Publ. No. 129, pp 263–269

OECD (2010) Paying for biodiversity: enhancing the cost-effectiveness of payments for ecosystem services. OECD Publishing, Paris,. https://doi.org/10.1787/9789264090279-en

Ostrom E (2010) Beyond markets and states: polycentric governance of complex economic systems. Am Econ Rev 100

Ottoni MV, Ottoni Filho TB, Lopes-Assad MLRC, Rotunno Filho OC (2019) Pedotransfer functions for saturated hydraulic conductivity using a database with temperate and tropical climate soils. J Hydrol 575:1345–1359

Ottoni MV, Ottoni Filho TB, Schaap MG, Lopes-Assad MLRC, Rotunno Filho OC (2018) Hydrophysical database for Brazilian soils (HYBRAS) and pedotransfer functions for water retention. Vadose Zone J 17(1):1–17

Parrota JA, Turnbull JW, Jones N (1997) Catalyzing native forest regeneration on degraded tropical lands. For Ecol Manage 99(1,2):21–42

Rao AR, Bhattacharya D (1999) Comparison of Hurst exponent estimates in hydrometeorological time series. J Hydrol 4:225–231

Rebello VPA, Getirana ACV, Lakshimi V, Rotunno Filho OC (2016). Monitoring drought in Brazil by remote sensing, Chapter 20. In: Lakshmi V (ed) Remote sensing of hydrological extremes. Springer International Publishing Switzerland 2016, pp 197–218, 251p

Renar KG, Foster GR, Weesies GA, Mccool DK, Yoder DC (1997) Predicting soil erosion by water: a guide to conservation planning with the revised universal soil loss equation. Agricultural Handbook, no 703. Agricultural Research Service, USDA, Washington, DC

Rosário LS (2013) Análise das áreas de proteção permanente com suporte de geotecnologias face a possíveis alterações no código florestal brasileiro – estudo de caso: bacia do rio Piabanha/RJ (Water quality spatial modeling applied to Piabanha river watershed/RJ). M.Sc. Thesis, Civil Engineering Program, COPPE/UFRJ, Rio de Janeiro, RJ, Brasil, 162 p

Rotunno Filho OC, Soulis ED, Kouwen N, Abdeh-Kolahchi A, Pultz TJ, Crevier Y (1996) Soil moisture in pasture fields using ERS-1 SAR data: preliminary results. Can J Remote Sens 22(1):95–107

Rotunno Filho OC, Araujo AAM, Xavier LNR, Moreira DM, Di Bello RC, Xavier AE, Araujo LMN (2014) Soil moisture and soil water storage using hydrological modeling and remote sensing. Chapter 14. In: Teixeira WG et al. (eds) Applications of soil physics in environmental analyses: measuring, modelling and data integration, progress in soil science. Springer International Publishing Switzerland, pp 307–345, 499 p. https://doi.org/10.1007/978-3-319-06013-2_14

Rotunno Filho OC, Paiva K, Rebello VPA, Araujo AAM (2019) How to integrate ground-truth and satellite image to estimate surface soil moisture at the field and watershed scales? In: 16th international symposium on water-rock interaction and 13th applied isotope geochemistry (1st IAGC), 2019, Tomsk–Russia, pp 1–5

Rotunno Filho OC (1989) Comparação de dois métodos de otimização em modelos hidrológicos do tipo chuva-vazão (Comparison of two optimization methods in rainfall-runoff models). M.Sc. Thesis, Civil Engineering Program, COPPE/UFRJ, Rio de Janeiro, RJ, Brasil, 176 p

Rotunno Filho OC (1995). Soil moisture mapping using remote sensing and geostatistics applied to rainfall-runoff models. Ph.D. thesis, Department of Civil Engineering, University of Waterloo, Canada, 396 p

Salles D, Barraque B, Busca D, Garin P (2006) L'eau des villes et l'eau des champs. Négociation territoriale et génie de l'environnement, Technical Report

Salzman J (2005) Creating markets for ecosystem services: notes from the field. N Y Univ Law Rev 80:870–962

Santos P, Brito B, Maschietto F, Osório G, Monzoni M (2012) Marco regulatório sobre pagamento por serviços ambientais no Brasil. Imazon; FGV. CVces, Belém

Schomers S, Matzdorf B (2013) Payments for ecosystem services: a review and comparison of developing and industrialized countries. Ecosyst Serv 6:16–30. https://doi.org/10.1016/j.ecoser.2013.01.002

Seppelt R, Dormann C, Eppink F, Lautenbach S, Schmidt S (2011) A quantitative review of ecosystem service studies: approaches, shortcomings and the road ahead. J Appl Ecol 48:630–636

Silva JS, Calmant S, Seyler F, Rotunno Filho OC, Cochonneau G, Mansur WJ (2010) Water levels in the Amazon basin derived from the ERS-2 and ENVISAT radar altimetry missions. Remote Sens Environ 114:2160–2181

Silva JS, Seyler F, Rotunno Filho OC, Calmant S, Roux E, Araujo AAM (2012) Water levels dynamics of Amazon wetlands at the watershed scale by satellite altimetry. Int J Remote Sens 33:3323–3333

Silva FP, Justi da Silva MGA, Rotunno Filho OC, Pires GD, Sampaio RJ, Araujo AAM (2018) Synoptic thermodynamic and dynamic patterns associated with Quitandinha river flooding events in Petropolis, Rio de Janeiro (Brazil). Meteorol Atmos Phys 131(4):845–862. https://doi.org/10.1007/s00703-018-0609-2

Silva FP, Rotunno Filho OC, Justi da Silva MGA, Pires GD, Sampaio RJ, Araujo AAM (2019) Observed and estimated atmospheric thermodynamic instability using radiosonde observations over the city of Rio de Janeiro, Brazil. Meteorol Atmos Phys 132:297–314. https://doi.org/10.1007/s00703-018-0609-2
Siska EM, Takara K (2015) Achieving water security in global change: dealing with associated risk in water investment. Procedia Environ Sci 28:743–749
Soares A, Leão M (2015) Contaminação dos Mananciais por Micropoluentes e a Precária Remoção desses Contaminantes nos Tratamentos Convencionais de Água para Potabilização. Doutrina Internacional 14(24):36–85
Sete Soluções e Tecnologia Ambiental (2017) Estimativa de custos de recuperação de áreas degradadas, Technical report
Sorooshian S, Arfi F (1982) Response surface parameter sensitivity analysis methods for postcalibration studies. Water Resour Res 18:1531–1538
Sorooshian S, Dracup JA (1980) Stochastic parameter estimation procedures for hydrologic rainfall-runoff models: correlated and heterocedastic error cases. Water Resour Res 16:430–442
Sorooshian S, Gupta VK (1985) The analysis of structural identifiability: theory and application to conceptual rainfall-runoff models. Water Resour Res 21(4):487–495
Sorooshian S, Gupta VK, Fulton JC (1983) Evaluation of maximum likelihood parameter estimation techniques for conceptual rainfall-runoff models: influence of calibration data variability and length on model credibility. Water Resour Res 10:251–259
Souza VAS, Moreira DM, Rotunno Filho OC, Rudke AP (2020) Extreme rainfalls events in Amazonia: the Madeira river basin. Remote Sensing Appl Soc Environ 18:100316
Speed R, Yuanyuan L, Le Quesne T, Pegram G, Zhiwei Z (2013) Basin water allocation planning: principles, procedures and approaches for basin allocation planning. Asian Development Bank, GIWP, UNESCO, and WWF-UK, 144 p
Srinivasan V, Konar M, Sivapalan M (2017) A dynamic framework for water security. Water Secur J 1:12–20
Superti E, Auberti C (2015) Pagamentos por Serviços Ambientais na Amazônia: o desvio de um conceito – casos do Amapá e Acre. Desenvolvimento e Meio Ambiente 35:209–224
Tejeiro G (2013) Os sistemas de pagamento por serviços ambientais como ferramenta de gestão dos riscos decorrentes das mudanças do clima. In: Lavratti P, Tejeiro G (Org) Direito e mudanças climáticas: Pagamento por Serviços Ambientais, fundamentos e principais aspectos jurídicos. São Paulo, Instituto O Direito por um Planeta Verde, pp 46–97
Torrence C, Compo GP (1998) A practical guide to wavelet analysis. Bull Am Meteor Soc 79:61–78
Treistman F, Maceira MEP, Penna DDJ, Damázio JM, Rotunno Filho OC (2020) Synthetic scenario generation of monthly streamflows conditioned to the El-Niño-Southern Oscillation: application to operation planning of hydrothermal systems. Stoch Env Res Risk Assess 1:1–23
Troutman BM (1982) An analysis of input errors in precipitation-runoff models using regression with errors in the independent variables. Water Resour Res 18:947–964
Troutman BM (1985a) Errors and parameter estimation in precipitation-runoff modelling, 1, Theory. Water Resourc Res 21:1195–1213
Troutman BM (1985b) Errors and parameter estimation in precipitation-runoff modelling, 2, A case study. Water Resour Res 21:1214–1222
Troy TJ, Konar M, Srinivasan V, Thompson S (2015) Moving socio-hydrology forward: a synthesis across studies. Hydrol Earth Syst Sci 19:3667–3679. www.hydrol-earth-syst-sci.net/19/3667/2015/. https://doi.org/10.5194/hess-19-3667-2015
Tucker C, Pinzon JE, Brown E, Slayback DA, Pak EW, Mahoney R, Vermote EF, El Saleous N (2005) An extended AVHRR-8km NDVI dataset compatible with MODIS and SPOT vegetation NDVI data. Int J Remote Sens 26(20):4485–4498
Vásquez ILP, Araujo LMN, Molion LCB, Abdalad MA, Moreira DM, Sanchez A, Barbosa HA, Rotunno Filho OC (2017) Historical analysis of interannual rainfall variability and trends in southeastern Brazil based on observational and remotely sensed data. Clim Dyn. https://doi.org/10.1007/s00382-017-2642-9

Vásquez ILP (2018) Análise estocástica e identificação de padrões de variabilidade espaço-temporal da precipitação no sudeste brasileiro e Amazônia (Stochastic analysis and identification of space-temporal variability patterns of precipitation in southeast Brazil and Amazonia). D.Sc. Thesis, Civil Engineering Program, COPPE/UFRJ, Rio de Janeiro, RJ, Brasil, 256p

Villarini G, Smith JA, Baeck ML, Vitolo R, Stephenson DB, Krajewski WF (2011) On the frequency of heavy rainfall for the midwest of the United States. J Hydrol 400(1–2):103–120

Vital D, Resck S (2002) A conservação da água via Terraceamento em Sistemas de Plantio Direto e Convencional do Serrado, Circular Técnica 22, Embrapa, Ministério da Agricultura, Pecuária e Abastecimento, Planaltina, Distrito Federal

Vörösmarty CJ, McIntyre PB, Gessner MO, Dudgeon D, Prusevich A, Green P, Glidden S, Bunn SE, Sullivan CA, Liermann CR, Davies PM (2010) Global threats to human water security and river biodiversity. Nature 467. https://doi.org/10.1038/nature09440

Wunder S (2005) Payments for environmental services: some nuts and bolts, vol 42. Center for International Forestry Research, Jakarta, 24 p

Wjuniski BS, Fernandez RG (2010) Karl Polanyi and us: the contemporary significance of Polanyi´s thought. Braz J Political Econ 30(119):420–437

Xavier AE (2001) Hyperbolic penalty: a new method for nonlinear programming with inequalities. Int Trans Oper Res 8:659–671

Xavier AE (2010) The hyperbolic smoothing clustering method. Pattern Recogn 43:731–737

Xavier AE, de Oliveira AAF (2005) Optimal covering of plane domains by circles via hyperbolic smoothing. J Global Optim 31:493–504

Xavier LNR, Becker M, Cazenave A, Longuevergne L, Llovel W, Rotunno Filho OC (2010) Inter-annual variability in water storage over 2003–2008 in the Amazon Basin from GRACE space gravimetry in situ river level and precipitation data. Remote Sens Environ 114:1629–1637

Xavier LNR, Araujo AAM, Rotunno Filho OC (2005) Bayesian kriging and GLUE applied to estimation of uncertatinty due to precipitation representation in hydrological modelling. Predicition in Ungaged Basins: Promises and Progress (Proceedings of Sumposium S7 held during Seventh IAHS Scientific Assembly at Foz do Iguaçu, Brasil). IAHS Publ., p 303

Xavier AE, Rotunno Filho OC, Canedo de Magalhães P (2022) The hyperbolic smoothing approach for automatic calibration of rainfall-runoff models. World Acad Sci Eng Technol Int J Environ Ecol Eng 14(3):37–41

Xavier LNR (2012) Modelagem hidrológica com o aporte de dados da missão espacial GRACE: aplicação a bacias brasileiras (Hydrological modeling with GRACE data: application to Brazilian watersheds). D.Sc. Thesis, Civil Engineering Program, COPPE/UFRJ, Rio de Janeiro, RJ, Brasil, 250 p

Yapo PO, Gupta HV, Sorooshian S (1998) Multi-objective global optimization for hydrological models. J Hydrol 204:83–97

ZEE-RJ (2008) Uso e cobertura do solo do estado do Rio de Janeiro (escala 1:100.000). Disponível em http://www.rj.gov.br/web/sea/exibeConteudo?article-id=282959. Acesso em 20 de março de 2012

Paths to Clean Water Under Rapidly Changing Environment

Natalia Guseva and Oleg Savichev

1 Introduction

The surface area of the Earth is 510.2·10^6 km^2, with the World Ocean accounting for 70.8% (https://bigenc.ru/geography/text/5310930). However, the World Ocean is the largest but not the only reservoir of water on earth. According to various data, it contains 94.0–96.5% of all of Earth's water (Mikhaylov et. al., 2005; Hendrics, 2010), while the remaining water is locked up in glaciers, lakes, rivers, water reservoirs, bogs, and subsurface.

These estimates are certainly rough and will be refined. At the same time, it is obvious that all waters on the Earth are interconnected due to their physical nature. Therefore, significant changes in at least one of water bodies, apart from the World Ocean, would definitely affect the entire hydrosphere of the Earth. This also concerns a large part of land in the Northern Eurasia within the administrative boundaries of the Russian Federation. There are significant resources of surface water and groundwater which are not typical for the main part of land in Eurasia, North and South America, Africa, and Australia. These enormous water resources are confined to the vast territories of permafrost that could be compared to Antarctica, the Arctic Ocean islands, and the arctic part of North America. In addition, there is a huge West Siberian wetland which, on the one hand, is similar to that in Amazonia, but, on the other hand, is different in terms of specific conditions and properties.

In this context, the present study is aimed at the analysis of:

(1) Current situation and water pollution in the Russian Federation (including the overview of water resources, chemical composition, and quality of surface water and groundwater);

N. Guseva (✉) · O. Savichev
Department of Geology, National Research Tomsk Polytechnic University, 30, Lenin Ave, 340050 Tomsk, Russia
e-mail: guseva24@yandex.ru

Y. Li et al. (eds.), *BRICS Countries: Sustainable Water Resource Management and Pollution Control*, https://doi.org/10.1007/978-981-99-9581-3_2

(2) Challenges related to water pollution (such as eutrophication of water bodies, water logging, permafrost landscape transformation, point and nonpoint pollution sources) and possible ways to address them via engineering (various treatment technologies) and management practices (water policy: administrative and economic regulations);
(3) Changes in water quality due to rapidly changing environment and with regard to ongoing water resource protection and management.

The present study is focused on the above-mentioned aim and objectives. The obtained results do not fully represent the topic in question, but they provide a thorough overview of the issues related to water security and environmental concern in the early twenty-first century in the Russian Federation.

2 Water Pollution in Russia: Current Situation

2.1 *Overview of Russia's Water Resources*

The Russian Federation (Russia) covers an enormous territory in eastern Europe and northern Asia (European and Asian Russia, respectively). The climates and landscapes change across the country from the Arctic, subarctic, and temperate zones to subtropics and from Arctic deserts and tundra to Mediterranean forests through tundra forests, taiga, deciduous forests, wooded steppe, steppe, semideserts, and deserts (Reymers, 1990). Russia is bordered by seas of the Atlantic, Arctic, and Pacific oceans. They are, namely, the Baltic (21,500 km^3), Black (555,000 km^3), and Azov (290 km^3) seas of the Atlantic basin; the Barents (316,000 km^3), White (6000 km^3), Kara (98,000 km^3), Laptev (353,000 km^3), East Siberian (49,000 km^3), and Chukchi (42,000 km^3) seas of the Arctic basin; and Bering (3796,000 km^3), Okhotsk (1316,000 km^3) and Japan (1631,000 km^3) seas of the Pacific basin. The inland waters include a large freshwater Caspian Sea of 78,000 km^3 (Shilkomanov, 2008; Gusev, 2019; Government Report, 2019).

Russia possesses extremely rich resources of surface waters and groundwaters (Water Code, 2018). The surface water resources comprise seas or their parts; rivers, creeks, and canals; natural and manmade lakes (water reservoirs or ponds, <1000,000 m^3); flooded quarries; wetlands; springs and geysers; glaciers and permanent snow fields. The groundwaters occur in artesian basins and aquifers. Russia's static water resources total approximately 88,900 km^3 of fresh water, while the long-term mean annual renewable amount is approximately 4200–4300 km^3/yr, of which ~95% of streams are sourced within Russia (Government Report, 2019).

Rivers over the greatest part of the country belong to the Arctic basin, including the catchments of Severnaya Dvina, Pechora, Onega, Mezen, etc., in European Russia, and Ob, Yenisei, Lena, Taz, Pur, Indigirka, Kolyma, Yana, etc. in Asian Russia. Several rivers flow into inland seas: Volga, Ural, Terek, Sulak, and Samur. The Atlantic basin receives waters of the Don, Kuban, Dniepr, Zapadnaya Dvina, Neva,

Neman, and some other rivers. The Amur, Kamchatka, Anadyr, Tym, etc. catchments belong to the Pacific basin. The largest river catchments are Ob' (2990,000 km^2), Yenisei (2619,000 km^2), Lena (2478,000 km^2), Amur (2050,000 km^2), and Volga (1380,000 km^2).

The river flow has changed from 3700 to 4700 km^3/yr over the period of large-scale systematic hydrological surveys since the 1930s. According to the Government Report (2019), the volume of lakes and reservoirs exceeds 4000 km^3, and the total surface area of water bodies reaches 344,480 km^2. The active storage of water reservoirs used directly for flow regulation is 324 km^3 (Shilkomanov, 2008) and makes up 790 km^3 together with the passive storage, which is used for head pressure, alluvium accumulation, and sanitary purposes but is not intended for flow regulation. The inferred (potentially useful) groundwater flow reaches 317.65 km^3/yr, including 27.97 km^3/yr of measured flow. Wetlands cover 1527,800 km^2, and the swamp water resources are at least 1500 km^3/yr (Shilkomanov, 2008; Gusev, 2019; Government Report, 2019).

Currently no more than 2% of resources are stored in different water bodies and 3–4% of inferred resources have been in use (Tables 1 and 2). In spite of high total water availability throughout the country, some regions suffer shortage of good quality water. Water scarcity may have several reasons: uneven distribution of water over the territory (Table 3); poor coordination between water demand and supply; run-off variability; insufficient water quality (below the standards) (Government Report, 2019). For instance, the territory of the Caspian, Black, and Azov basins accommodates about 75% of Russia's population and 80% of industry and farming but has only 16% of the surface water resources (Government Report, 2019).

More than 70% of the used waters are returned to surface water bodies (Table 2) as effluents treated to different degrees, but some still contain pollutants (Table 4).

2.2 *Overview of Water Composition and Quality in Russia*

According to the State Standards (2001, 2003), water chemistry is an assemblage of substances that are present in water in different chemical and physical states; the quality of water refers to its chemical, physical, and biological properties with respect to suitability to particular intended uses according to the respective standards. The water quality standards practiced in Russia are different for potable and non-potable water for domestic (drinking and household) and fishing or industrial and commercial purposes, respectively. Correspondingly, the concept of maximum contaminant level (MCL) is applied to concentrations of substances in non-potable (MCL_{np}) and potable (MCL_p) water.

Seawater chemistry is different in deep, shallow, and transitional water zones. Deep water chemistry is quite stable and generally represents the global budget, while the chemistry of shallow and transitional waters depends on the balance of water, heat, and chemical elements, including inputs of material and energy with river run-off (Alekin, 1970; Lisitsyn, 1994).

Table 1 Water resources of Russia and their use (Shilkomanov, 2008; Gusev, 2019; State Report, 2019)

Water resources				Units	Value
Long-term data since the 1930s	Average			km^3/yr	4250.1
	Maximum			km^3/yr	4746.0 (1974)
	Minimum			km^3/yr	3758.0 (1954)
Groundwater as in 2018	Inferred resources			km^3/yr	317.65
	Measured resources (reserves)			km^3/yr	27.97
25 largest natural and manmade* lakes (without Caspian sea)				km^3	4140.75
Ground ice				km^3	17,178
Glaciers (conservative estimate)				km^3	13,601
Wetlands (conservative estimate)				km^3/yr	1520
Supply from natural sources	Total			km^3/yr	68.02
	From	Rivers and lakes	For use	km^3/yr	44.90
			For transfer	km^3/yr	6.25
		Seas		km^3/yr	5.46
		Aquifers		km^3/yr	11.41
Used water from natural bodies				km^3/yr	52.93
Discharged water	Into natural water bodies	Surface waters	Total	km^3/yr	48.94
			Transferred	km^3/yr	8.91
		Groundwaters		km^3/yr	0.36
	Into canals			km^3/yr	–
	Into settling ponds, depressions, and irrigation reservoirs			km^3/yr	0.61

Note Reserves of natural and manmade lakes include: Ladoga, Onega, Ilmen, Baikal, and Khanka lakes; Ivankovo, Uglich, Rybinsk, Gorky, Cheboksary, Kuibyshev, Saratov, Volgograd, Iricla, Tsimlansk, Krasnodar, Kama, Votkinsk, Novosibirsk, Sayan-Shusha, Krasnoyarsk, and Irkutsk reservoirs

The chemistry of inland waters is controlled by slope, surface, and groundwater run-off components. The contents of dissolved inorganic compounds normally decrease with increasing flow in the conditions of pronounced streams. Time-dependent variability of water chemistry and the probability for elements to reach maximum concentrations are generally proportional to the long-term annual and seasonal run-off variability. The contents of dissolved substances in rivers, lakes, wetlands, and aquifers have multiple controls: availability of agents for reactions; sorption capacity of subaerial sediments, aquifers, and underlying rocks; nominally equilibrium contents of dissolved substances; rates of dissolution–precipitation and sorption–desorption processes in specific landscapes and ecosystems. All these factors depend on turnover rates (Shvartsev, 2008; Savichev and Guseva, 2018).

Table 2 Average water use in Russia over 2015–2018 (Government Report, 2019), km^3

No.	Parameter	Value
1	Supply from natural water bodies (including seawater), total	68.8
1.1	Supply for use	61.0
1.1.1	Surface waters	51.7
1.1.2	Groundwater	9.3
2	Used fresh water, total	54.0
2.1	Drinking and service water	7.9
2.2	Process water	30.5
2.3	Irrigation, stockwater development, and farming	7.1
3	Recycled and reused water, total	139.9
4	Losses on transport	6.9
5	Discharge into surface waters, without surplus spill	42.2
5.1	Effluent pollution	14.0
5.1.1	Untreated	2.9
5.1.2	Poorly treated	11.1
5.2	Water treated to effluent quality	26.2
5.3	Water treated to standard quality	2.0

As a result, the concentrations of contaminants in surface waters and groundwaters throughout the Russian territory may exceed the maximum level at least in some periods of a hydrological year. Therefore, the above-MCL concentrations or low water quality according to some MCL-based standards not necessarily would mean pollution by wastewater discharge or any human activity.

For instance, 25% of the territory in the Tomsk and Tyumen' regions (including the Yamal-Nenets and Khanty-Mansi districts) in West Siberia is swampy (Savichev, 2010a). Consequently, surface waters and groundwaters in the upper hydrogeodynamic stage may, to a high probability, have pH < 5–6 and contain more than 15 mg/l organic matter (according to chemical oxygen demand, COD), or hydrocarbons, phenols, products of organic decay (e.g., NH_4^+), as well as compounds of organic acids and metals, especially iron. Some manmade and natural effects on water chemistry are considered in more detail below.

The water quality estimates are reported in this chapter as determined by the methods practiced in Russia. Specifically, the suitability of water for domestic uses is estimated on the basis of an additive effect factor (1) and the general quality of surface waters is judged according to the probability that contaminants exceed the maximum level (Water Pollution, 2002):

$$\sum_{1}^{N} \frac{C_i}{LC_i} \leq 1 \tag{1}$$

Table 3 Long-term average river flow in Russia, according to sea basins (Government Report, 2019)

Sea basin	River	Catchment area, km^2	River length, km	Average flow, km^3/yr	Dry year flow (P 95%), km^3/yr	Water availability, × 10^3 m^3/yr	
						Per 1 km^2	Per person
Baltic	Neva	189.7	750	83	57.1	417.7	11.7
	Zapadnaya Dvina	88	1020	20.3	11.6	206.9	20.2
White	Severnaya Dvina	357	750	103	58	225.8	46.2
Barents	Pechora	322	1814	131	99	403.7	179.3
Azov	Don	422	1870	26.8	11.7	66.1	1.2
	Kuban	57.9	970	14.4	8.6	139.5	2.8
Caspian	Terek	37.4	–	11.1	8.75	255.7	4
	Ural	236	2530	8.87	3.3	33.4	6.2
	Volga	1380	3694	260	182.9	175	4
Kara	Yenisei (+Angara)	2580	3844	651	528	244.2	221.5
	Ob (+Irtysh)	2990	3676	407	282	178.6	54.4
Laptev	Lena	2490	4337	543	424	209.2	446.7
East Siberian	Coastal rivers	1097.6	–	255	204	232.3	952.5
Bering	Coastal rivers	472.3	–	312	249	660.6	2331.8
Okhotsk	Amur	1855	2855	359	209	185	88.7
Japan	Coastal rivers	130.3	–	37.4	29.2	287.1	21.7

where C_i and LC_i are, respectively, actual and maximum levels of the i-th contaminant of hazard grades 1, 2; N is the number of analyzed substances in water, for which MCL have been established by Sanitary Norms and Regulations (2000, 2001, 2002); the quality estimate is yes/no (suitable/unsuitable).

The total contents of dissolved solids and water chemistry are classified (Alekin, 1970) according to TDS or predominant major ions. The TDS (salinity) types of water are: fresh, very low salinity, < 0.1 g/L (‰); fresh, low salinity, 0.1–0.2 g/L (‰); fresh, medium salinity, 0.2–0.5 g/L (‰); fresh, elevated salinity, 0.5–1.0 g/L (‰); brackish, 1–25 ‰; seawater-level salinity, 25–50 ‰; brines, > 50 ‰. The major-ion types are: C (HCO_3^-), S (SO_4^{2-}), and Cl (Cl^-); Ca^{2+}, Mg^{2+}, or Na^+ according to anions, cations, and their relative contents. Waters with dichromate values > 15 mg/L COD and permanganate values > 5 mg/L COD are classified as organic-rich. The pH types are: strongly acidic (pH < 3), acidic (pH = 3–5), subacidic (pH 5–6.5),

Table 4 Pollutants discharged into water bodies, average over 2015–2018 (Government Report, 2019), kt/yr

Pollutant	Value
Suspended matter	186.1
Dry residue	6787.6
Chlorides (Cl-)	5827.7
Sulfate anion ($SO_4{}^{2-}$)	1943.3
Calcium (Ca^{2+})	339.7
Magnesium (all soluble Mg species)	33.8
Sodium (Na +)	429.3
Potassium (K^+)	72.4
Nitrate anion ($NO_3{}^-$)	409.4
Nitrite anion ($NO_2{}^-$)	6.1
Ammonium	60.0
Phosphates (according to P)	15.0
Organic matter (according to COD)	310.7
Organic matter (according to BOD, total)	141.9
Carbamide	5.3
Fats/Oils (natural)	2.0
Petroleum	2.1
Phenol	0.017
Fluorine (F^-)	2.0
Iron (all soluble Fe^{2+} and Fe^{3+} species)	2.5
Nickel (Ni)	0.027
Manganese (Mn)	0.284
Copper (Cu)	0.035
Zink (Zn)	0.303
Lead (Pb)	0.005
Chromium (Cr), kg	0.016
Aluminium (Al), t	0.509

neutral (pH = 6.5–7.5), subalkaline (pH = 7.5–8.5), alkaline (pH – 8.5–9.5), and strongly alkaline (pH > 9.3) (Guseva, 2007).

2.2.1 Lakes

Water chemistry in natural and manmade lakes depends mainly on the lake origin, relative volumes of deep and shallow water, inflow and outflow, and landscape. Most of fresh lake water (95%) is stored in Lake Baikal (Shiklomanov, 2008). Water in large lakes is mostly fresh, with very low to medium average salinity; in the low-water season, the salinity increases to medium and elevated levels. The salinity is the lowest (<0.5 g/L TDS) in taiga flowing lakes, in the absence of groundwater recharge, and the highest (>50 g/L TDS) in non-flowing water bodies of steppe and semi-desert landscapes (Alekin, 1970).

The major-ion chemistry of water depends on climate and landscape conditions. The predominant ions are HCO_3^- and Ca^{2+} in plainland water bodies of forest landscapes in the temperate climate zone; SO_4^{2-} and Na^+ at higher TDS in wooded steppe landscapes; and SO_4^{2-}–Na^+, Cl^-–Na^+ or HCO_3^-–Na^+ in evaporating brackish or salt lakes of dry steppes and semi-deserts (Bogoslovsky, 1960; Alekin, 1970). As far as TDS values in surface waters increase, the contents of many substances that are normally limited to very low levels increase correspondingly up to economic amounts.

The TDS budget is maintained by both surface run-off and groundwaters. As it was shown for Itkul and Shira, a pair of lakes spaced at 4–5 km in Khakassia, with TDS values of 0.6–0.7 g/L and 12–31 g/L, respectively, the annual groundwater flow from Itkul to Shira reaches 6791,000 m^3/yr and the respective loss of dissolved solids is 35,697 t/yr. Correspondingly, the Shira water is more saline than Itkul where evaporation is less significant for water chemistry while the water–rock interaction is shorter. A similar mechanism of TDS forcing in lakes of arid climate zones was reported also from other areas of Northern and Central Asia (Guseva and Savichev, 2016).

Water bodies in highland and permafrost areas constantly or seasonally receive inputs of saline groundwater or even brines, including waters rich in heavy metals. Lakes in tundra, tundra forests, and taiga often have low pH values and organic carbon contents above 30–50 mg/L COD, and contain heavy metal compounds (Savichev et al., 2011).

These patterns are typical of both natural lakes and manmade dam lakes (reservoirs). The contents of some contaminants may increase additionally in the case of wastewater inputs from point or diffuse sources. The anthropogenic impact on water bodies, especially on small and/or non-flowing lakes, is generally heavier than on rivers which are prone to self-cleaning. In this respect, waters of many big lakes and reservoirs, which are largely used for water supply and discharge of effluents, are classified as moderately polluted and those of small lakes and reservoirs are moderately, heavily, and extremely polluted (Shiklomanov, 2008; Gusev, 2019).

2.2.2 Rivers

Russia's rivers in open-water periods are on average fresh, with low or moderate salinity and HCO_3^-–Ca^{2+} major-ion chemistry in the forest, tundra, and tundra forest landscapes; locally fresh, with very low salinity and sulfate chemistry, in taiga, tundra forests and tundra; and fresh, with elevated or high salinity (brackish) and HCO_3^-–Ca^{2+} or HCO_3^-–Na^+ chemistry, in wooded steppe and steppe zones.

The watersheds and valleys of taiga, tundra forest, and tundra rivers are often swampy and carry significant amounts of organic matter and products of its decay into the river network (Savichev et al., 2016, 2020). The valleys of many rivers in

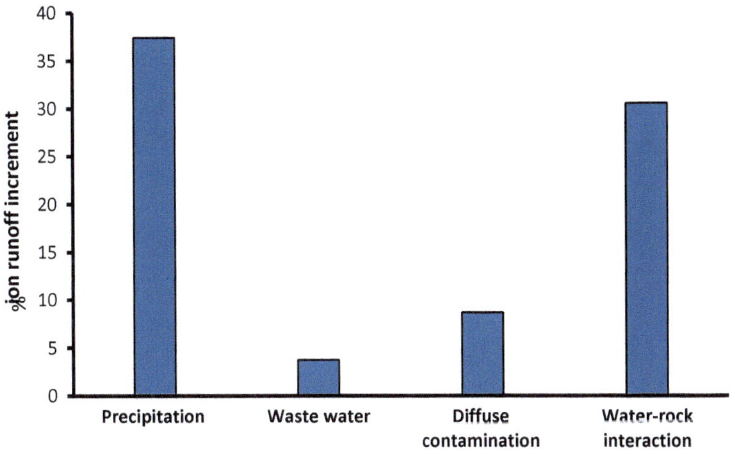

Fig. 1 Major ion sources versus ion runoff increment (cation sink is 7% of total ion sink; total 87%), middle Ob (Savichev, 2010b)

the steppe zone are likewise swampy, and the water is often enriched in organic acids and their compounds, and metals, especially iron. Correspondingly, the river water quality in such areas is often moderately polluted or polluted, with Fe, Cu, Zn, Al, hydrocarbons, phenols, nitrites, ammonia, and dichromate values above MCL (Shiklomanov, 2008; Gusev, 2019).

The high contents of contaminants are commonly attributed to man-caused impacts, but they don't always correlate with long-term production figures in industry and agriculture. Water chemistry in the middle reaches of the Ob River and its tributaries between the mouths of Tom and Irtysh (Middle Ob) was studied in detail in order to discriminate between natural and man-caused element loads (Fig. 1). The Middle Ob water chemistry was inferred to be mainly of natural origin in terms of major ions and organic compounds resistant to oxidation (high COD). The main natural run-off trend in this part of the catchment consists of gradual increase in organic carbon due to progressive swamping (Fig. 2).

The man-caused changes in ion sinks through 1970–2000s are mainly due to inputs of toxic and easily oxidizable (BOD-based) organic compounds, hydrocarbons, inorganic nitrogen compounds, and microorganisms into the river waters. The increasing inputs of contaminants are due mainly to air pollution and uncontrolled diffuse sources in urban areas (Savichev, 2010b). Similar conclusions can be obtained for other large rivers of Northern Eurasia.

2.2.3 Groundwater

Groundwaters in Russia belong to several provinces with different water chemistry conditions (Zektser, 2007; Government Report, 2019). As in the case of surface

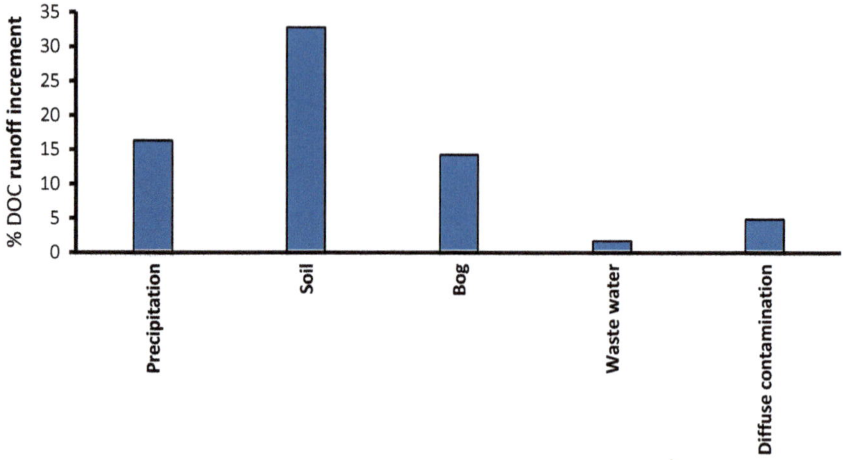

Fig. 2 DOC sources versus DOC runoff increment, middle Ob (total sink 70%) (Savichev, 2010b)

waters, the water chemistry patterns depend on climates and landscapes, as well as on stages of water–rock interaction and turnover rates which control the duration and surface area of the interaction (Shvartsev, 2008; Guseva, 2018).

Correspondingly, brackish and saline groundwaters and brines more often occur in wooded steppe, steppe, and semi-desert zones, and in permafrost areas. TDS values are the lowest in forest groundwaters from the upper stage (first aquifer from the surface, unconfined from above), which also often have the highest contents of DOC and decay products, as well as metals, specifically iron (Shvartsev, 2008; Government Report, 2019). Therefore, these groundwaters commonly have above-MCL Fe and NH_4^+ and high permanganate values throughout Russia (about 70% of the total area of polluted groundwater) (Government Report, 2019). High contents of Fe (Fe^{2+}), Si, Al, Mn, and some other elements also typically occur in artesian waters of different ages.

Groundwaters are markedly polluted in urban, industrial, farming, and mining areas, each having its specificity. Namely, high concentrations of hydrocarbons in oil and gas fields, or high organic compounds, NO_2^-, NH_4^+, phosphates, chlorides, and microorganisms in urban territories. Groundwater contaminants of hazard grades 1, 2 exceed MCL in about 20% of polluted areas (Government Report, 2019).

2.2.4 Marine Water

The chemistry and quality of marine water (salinity from 25–35 ‰ in deep sea to 0.5–1.0 ‰ at river inlets) depend on the percentage of river recharge relative to the sea volume, as well as on the density of population and industry in the basin. Correspondingly, the concentrations of petroleum, phenols, Cl-organic compounds, and heavy metals are above MCL in shallow marine waters along the coast and near river deltas, in areas of petroleum production (seas of Kara, Caspian, and partly Barents), as well as in most developed industrial regions (Baltic, Black, Azov, and Japan seas) (Bryzgalo et al., 2015; Government Report, 2019; Gusev, 2019).

3 Challenges and Opportunities to Treat Water Pollution

3.1 Eutrophication, Swamping, and Landscape Changes in Permafrost

Many water bodies undergo eutrophication which means that they increase bioproductivity by overly enrichment with minerals and nutrients, often coming from manmade sources, or by swamping. Anthropogenic eutrophication affects waters in areas of warm and wet climates in southern, central, and western European Russia, and in southern Urals, as well as small lakes and streams within urban areas throughout the country.

Other territories, such as West Siberia and northern European Russia, rather face challenges of swamping and landscape change driven by natural and manmade mechanisms in permafrost areas (northern ER), Western, Central, and Eastern Siberia (including the Pacific region).

Swamping occurs most often either as successive changes in eutrophic lake → eutrophic bog → mesotrophic bog → oligotrophic bog, or as water logging of dry forests at the boundary of mesotrophic and oligotrophic bogs because different amounts of water can pass through forest and wetland ecosystems. The former way of swamping was predominant in the first half of the Holocene, when vast poorly drained spaces in West Siberia were liberated from ice after the Last Glacial and became inundated by meltwater (Liss et al., 2001; Maloletko, 2008).

I–Watershed Elevation in the Baltic System; II–Levels of Swamp (River) Waters; III–Wetland Bed Elevation (Savichev and Paromov, 2014).

As the valley glaciers degraded, the run-off reduced, and the river network acquired its present form with numerous abandoned channels and oxbow lakes. The oxbow lakes were the first to experience eutrophication and became the centers of swamping. The peat thickness was growing at 0.7–1.0 mm/yr and has reached 11 m (3–4 m on average) (Liss et al., 2001; Pologova, Lapshina, 2002). As swamping progresses, the primary drainage network degrades and gives way to a new wetland system of lakes and streams. Some paleo watersheds become an underflow part of new streams as paleo river valleys are fully peatified. Swamping affects even medium rivers (Figs. 3 and 4).

The future evolution trend of wetlands in West Siberia remains unclear. According to some models, climate change may lead to increase in the number of lakes in northern Siberia (Shikolmanov et al., 2013). Another possibility is that water logging and run-off obstruction in the case of economic development in wetland territories may be favorable for the formation of bog lakes. On the other hand, forests of different

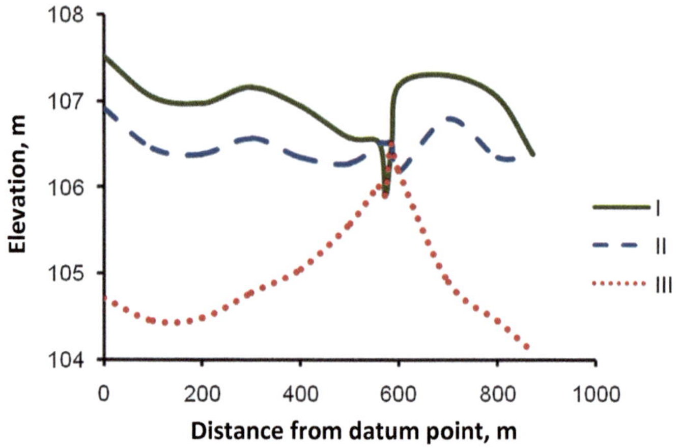

Fig. 3 Valley profile of a secondary wetland stream in the upper reaches of Trom-Yugan river, a tributary of the Ob (Khanty-Mansi district)

Fig. 4 Swampy valley of Parabel river, a tributary of the Ob (Tomsk region), without channel run-off over the greatest part of the year (dash line) (Savichev, 2008))

ages in peatbogs are vulnerable to spontaneous ignition and fire hazard, especially old forests of 100–120 years; such fires can last for years. At any scenario, wetlands that occupy hundreds and thousands of square kilometers cause negative effects on water quality in rivers, lakes, and aquifers that receive organic compounds, NO_2^-, NH_4^+, and Fe. Oxidation of organic matter consumes large amounts of dissolved oxygen, which causes death of fish in lakes and rivers during the low-water period (Savichev, 2010b).

Permafrost areas have particular water quality patterns in surface water bodies and aquifers. The topsoil that acts as heat insulation for permafrost can degrade upon climate warming and/or human loads. The ensuing erosion produces gullies that grow deeper at more than 1 m/yr and causes deformation of the watershed surface. The erosion favors tens of meters annual transfer of water accumulated in topographic lows, as well as flow of intrapermafrost saline water and brines.

3.2 Pollutant Sources of Surface Waters and Groundwaters

The sources of water pollution are either point (focused) or non-point (diffuse). Point sources have one identifiable cause of the pollution and are most often controllable. The maximum level of discharged contaminant substances and microorganisms is established at least every five years for each source. Monitoring and mitigation measures, such as water saving at enterprises or improving water treatment, are scheduled accordingly in order to reduce negative impacts on the receiving waters. If the measures are inefficient, the government water management authorities may forbid discharge of wastewaters or even close the enterprise.

The total amount of pollutant sinks in water bodies (Table 4) is not very high relative to the total river run-off (Fig. 1), but the discharge is commonly localized within limited sites and is received by small streams. This produces local impacts that are restricted to distances of 4–5 km and are detectable within 8 12 km downstream in medium and large rivers (Fig. 5 (Savichev, 2010b)). These estimates are consistent with data from elsewhere worldwide.

In addition to the controllable point sources, much pollution comes from diffuse non-point sources, such as run-off from urban areas, roads, and industrial facilities dispersed over large territories. The amount of diffuse pollution is assessed according to special regulations (Effluent Standards, 2018), but further theoretical and methodological work is required in this field.

The uncontrolled anthropogenic effects on water chemistry evaluated for the Middle Ob basin show up especially as high inputs of organic matter (high COD), petroleum, N, and P. The relative inputs of major ions, Fe, Cu, and DOC from urban areas are within 15% of all water transport in the Ob, but the absolute amount of contaminants exceeds markedly the total from point sources. Furthermore, the pollution from non-point sources is more heterogeneous spatially and affects considerably the chemistry of receiving rivers and groundwaters due to large inputs of N, P, and DOC.

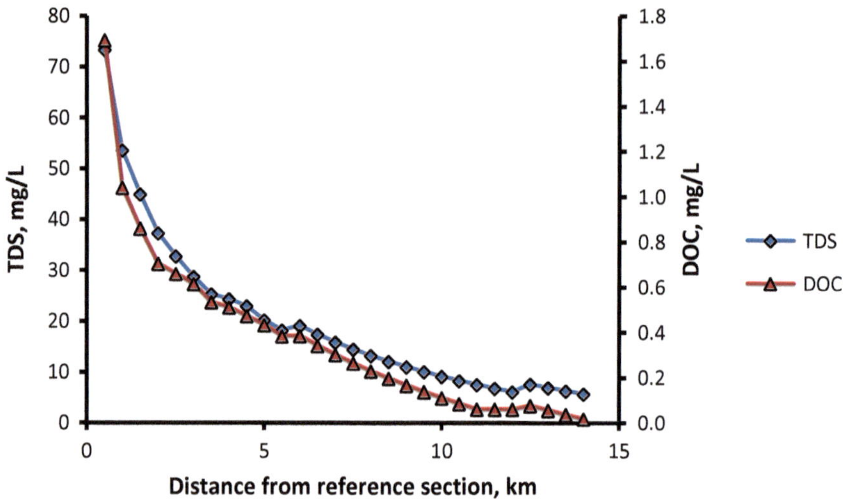

Fig. 5 Average deviation of TDS and DOC from the background as a function of distance from reference Sect. 0.5 km upstream of discharge at mouths of strongly polluted small rivers, Tom river (Savichev, 2010b)

3.3 *Water Treatment: Challenges and Opportunities*

The strategy of reducing human impacts on water bodies in Russia consists of several steps.

1. Water quality monitoring in natural water bodies and at industrial sites during feasibility studies and operation. The water quality in natural surface waters and groundwaters is monitored, respectively, by the federal agencies of Weather and Environment (*Rosgidromet*) and Subsoil Use (*Rosnedra*), two departments of the RF Ministry of Natural Resources and Environment. Additionally, systems for lake and river water supply and discharge are monitored by local water users and federal government supervision authorities. In industrial areas, water quality monitoring is performed during feasibility studies before construction and during the operation of production facilities, as part of regulatory measures for impacts on water bodies.
2. Establishing, reconciling, and approving effluent standards as to types and concentrations of pollutants and microorganisms, for periods of 1 to 5 years.
3. Mandatory annual reporting (form "2-TP-water use") on water supply, discharge of effluents, and contents of pollutants and microorganisms in wastewaters. The reported data are used to calculate water supply, effluent, and pollution charges. The charges depend on the amount and quality of effluents in which the contaminant contents may be within the regular standard approved for the respective period, above this standard but below the preliminary approved standard, or above the preliminary standard. The enterprises that discharge untreated wastewaters

pay charges factored relative to the charge for the level within the preliminary standard.

4. Imposing financial or administrative restrictions on water users; in the absence or poor efficiency of water treatment systems, the enterprises may be subject to closure (Effluent Standards, 2018). Installing and upgrading water treatment systems is a mandatory condition for wastewater discharge. Wastewater is treated through physical, chemical, physicochemical, and biological processes (Karpukhin, 2001; Water Treatment Regulations, 2012a, 2012b), which are chosen in each specific case according to the place, type, and toxicity of pollution. The choice aims at optimal practices for lowest costs, with regard to expenses for labor, material resources, power and fuels, and with reference to sanitary and fishery regulations (Water Treatment Regulations, 2012b; Wastewater Treatment, 2015a, 2015b; 2016).

The water treatment facilities are designed according to the properties and volume of wastewater, required water quality, methods for sludge treatment, and local conditions. Domestic sewage is treated at least by physical (settling and filtration) and biological methods and disinfected, while industrial waters require a full cycle of treatment by multiple processes.

The water treatment practices in Russia are facing a problem of water quality assessment using Eq. (2.1.2–1) (Savichev, 2010b). At $N > 20$ to 50, the water quality can be expected to approach almost pure H_2O, but the choice of normative parameters is ambiguous. Another problem is that the water treatment is costly and, by itself, may pose pollution risks to other environment components or deteriorate the conditions of nature use. Specifically, digestion and disposal of sludge consumes additional money and space (land) and pollutes air and water, even though the inputs from point and non-point sources are within a few percent of the total river run-off, at least in Asian and northern European Russia.

A team from the Tomsk Polytechnical University (Tomsk) studied the operation of a sewage treatment plant in Asian Russia and inferred that designing of such a facility, with a surrounding buffer recreation zone, should begin with inventory of the water and chemical budgets of the basin. The construction of treatment facilities is reasonable if untreated effluents exceed 5% of run-off at the basin outlet or if the effluents contain hazardous toxic elements and microorganisms (hazard classes 1, 2). An optimal trade-off between efficiency and costs can be achieved for a treatment plant that uses physical and biological cleaning with natural aeration and is located near natural topographic lows (river valleys or wetlands), provided that appropriate geotechnical measures of banking, storm and snowmelt draining, reinforcement and waterproofing, etc., are undertaken. Should the plant construction be found unreasonable, the effluents must comply with special requirements applicable to aquatic buffers for river valley segments flooded at 10% flood frequency (Savichev and Guseva, 2020).

Another important issue is to rehabilitate lands and water bodies that suffer from emergency discharges of wastewater or products (e.g., accidental oil spills from pipelines). Many accidents occurring yearly cause pollution to swamps, rivers, lakes, and aquifers, and those responsible for the emergency have to remediate the damage. Rehabilitation is mandatory after construction works and at the end of the designed operation life. In any case, rehabilitation should be part of design documents and include engineering (removal of coarse litter and landfilling) and biological (fertilizing and planting) measures. Rehabilitation is a mandatory part of any development project, but it is not always workable. For instance, applying the above strategy to remediation of a raised oligotrophic bog after an oil spill in West Siberia may be costly but eventually lead to the formation of a stable eutrophic bog. In this respect, rehabilitation is reasonable if the territory has no potential for self-recovery in 3–5 years and should follow the local natural conditions as close as possible. As proven by the available data from West Siberian taiga, rehabilitating a raised bog affected by a spill of 16–17 t of oil is inefficient.

At the final stage of a construction and rehabilitation project, the water and chemical budgets are to be re-inventoried. Then the standards for the safe amount and quality of effluents are to be established if the operation cycle is designed to include water treatment. The impact of effluents can be considered acceptable and local-scale if it is limited to 5% of total run-off at the basin outlet. Rehabilitation is considered successful if the chemical budgets before and after the disturbance are no more than 20% different. In the absence of pre-construction data from the site, a neighbor background area is used for reference.

3.4 Policy

3.4.1 Introduction of Water Policies, Laws, and Regulations Related to Water Pollution Control in Russia

The national policy in water resource management in Russia is regulated by federal laws, President's decrees, and statutes of the RF government and the federal, regional, or local authorities adopted within their competence (Government Report, 2019). The main principles of the water management policy are formulated in the Water Code (2018) and in the law of Subsoil Use (2011).

The approved management methods are grouped according to different indicators and are (i) intended for prevention or mitigation of pollution; (ii) direct or indirect, and (iii) government- or user-level (Environment Law, 2001, 2017). The division into the prevention and mitigation measures against pending or effectuated pollution, respectively, appears to be the most important. The other indicators are subsidiary and are used for lower classification levels.

The prevention measures, in their turn, are of three groups, including resource-saving technologies, as the environmental impact is proportional to the amount of resources in use; education and culture, which are not synonyms, and the former cannot help without the latter; creation of aquatic buffers, nature reserves, and water protection systems (setting back for bringing together); investment project management at different stages. The stages of investment projects are: pre-investment (being responsible: no action, no consequences); investment (being conscious when acting); and operation (keeping the operation attended, like a child).

Note that the investment management can be discussed either from the side of nature users or from the side of a state's government. A user has the rights and duties to ensure environmental safety and is supposed to take the appropriate measures such that the design, construction, and operation of facilities were in full compliance with the environmental legislation. A government, with participation of citizens, develops the environmental laws and regulations and supervises their enforcement. Specifically, all facilities are constructed according to designs which include a special section on the projected environment impact and its mitigation.

A report on assessment of environmental impacts from an operation has to specify the goal of the project and to provide a brief description of the operation; list uncertainties in possible impacts; estimate the extent and character of the impacts; describe measures for prevention and/or mitigation of possible impacts; plan safety and rehabilitation measures; propose programs for monitoring before, during, and after the development; prove feasibility of the environmental solutions.

The government agencies verify the design documents in terms of engineering and environmental issues (check whether the design is consistent with starting data, engineering conditions, and legislation) and monitor construction and operation in order to prevent, detect, and stop violation of environmental regulations.

Possible responses to environmental problems include amelioration, rehabilitation, geotechnical measures for cleaning water bodies and improving water quality, treatment of wastewaters and air emissions, as well as recycling and reusing industrial and domestic wastes.

3.4.2 Water Resource Management in Russia

According to the Water Code (2018), the federal and local executive authorities responsible for the management of water resources make up a single system that provides implementation of the national policy in the use and conservation of water bodies. The system of federal and local executive authorities that provide the management of water resources as to their use and conservation is established by the entities of the Russian Federation in accordance with the constitutional foundations, the general organizational principles of the executive authorities, and the Water Code. The management is effectuated by the government of the Russian Federation and by a specially authorized government agency which acts at the federal, basinal, and local management levels.

There are twenty basin provinces in Russia: Baltic Sea, Barents-White Sea, Dvina-Pechora, Dnieper, Don, Kuban, West Caspian, Upper Volga, Oka, Kama, Lower Volga, Ural, Upper Ob, Irtysh, Lower Ob, Angara-Baikal, Yenisei, Lena, Anadyr-Kolyma, and Amur (Water Code, 2018).

In general, management at the federal and partly basinal levels focuses on main guidelines in the water use policy and on their general monitoring, whereas the basinal and local management deals with specific measures and programs. The management of water resources in the RF territory is performed by the executive authorities of republics, regions, districts, federal-level cities, and capitals of autonomous entities.

3.4.3 Environmental Impact Assessment (EIA)

The assessment of environmental impacts is based on the concepts of natural, natural-anthropogenic, and anthropogenic objects. As stated in the Environment Law (2001, 2017), a natural object is a natural ecosystem or a landscape and its constituents that have preserved their natural properties; a natural-anthropogenic object is a natural object that has been changed by human activity, and/or a manmade object that has properties of a natural object and has recreational and security values; an anthropogenic object is a manmade object created for human social needs, which has no properties of a natural object.

The key approach implies conservation of natural objects, as far as it is possible at given social and economic conditions, and transformation of anthropogenic objects into natural-anthropogenic ones which are operated according to natural laws and thus become more stable against external effects and require less maintenance efforts. This approach can ensure sustainable development of the mankind and its adaptation to biospheric changes.

Note that conservation does not mean maintaining natural objects in a fixed state but rather providing conditions for their dynamic equilibrium, with due regard for natural transformation laws and the features of their ambiance. On the other hand, natural objects are not necessarily superior in quality relative to natural-anthropogenic or anthropogenic objects. Another key approach to the conservation of water bodies consists of planning, implementing, and supervising environmental safety measures at all stages of the investment cycle (Savichev, 2012).

3.4.4 Planning of Safe Water Use and Conservation of Resources

The measures for the use and conservation of water and other resources in the territory of Russia are of several groups and concern information support of planning (monitoring and inventory of natural resources); strategic planning at the federal, basinal, and local levels: planning and operation for efficient and safe use of water resources, including as part of national programs for the use, rehabilitation, and conservation of environment components; modeling of ultimate states and tactical planning at the federal, basinal, local, and user levels: regulation of impacts on water bodies and atmosphere, including MCL standards for effluents into water bodies and emissions into air, as well as wastes treatment standards.

These measures aim, respectively, at estimating the quality of environment components and heat and mass budgets. The approaches to water quality assessment are described in Sect. 3.1.2. Planning is undertaken at the levels of government and water users. The data on water supply and on discharge of effluents into water bodies and air emissions, pollution, and handling of wastes are reported annually to specially authorized management offices. The information is included into annual summary reports on the state of the environment in the territory of the Russian Federation as a whole and in its different entities (Government Report, 2019). The data make basis for pollution charging.

3.4.5 Pollution Charges

The protocols for calculation and collection of pollution charges within the Russian Federation are regulated by the respective document approved by decree N 255 of the RF government, on 3 March of 2017. According to this document, the charges are for polluting emissions into air by stationary sources and effluents into water objects, as well as for storage and disposal of wastes (Wastes Management, 2012). The charging is controlled by the Federal Natural Resource Agency and its territorial departments. The charges are imposed on legal persons and sole proprietors (promisors) that perform any economic or other activity that may cause pollution in the territory of the Russian Federation, on its continental shelf, and in exclusive economic zones, except for the promisors that perform any economic or other activity exclusively as polluting operations and belong to category IV.

The size of water pollution charges depends on whether the actual pollution exceeds the MCL values established for a period of 1 to 5 years. The charges for above-MCL pollution are additionally factored.

3.4.6 Challenges and Opportunities for Russia's Water Policy to Water Pollution Control

The water use legislation of the Russian Federation is being progressively improving with reference to the global trends. However, there remain some weak points concerning the ownership of water bodies and some controversy in terminology. For instance, the concept of wetlands is poorly defined, even among researchers in Russia and abroad. Wetlands are considered in different geoscience contexts as surface or underground water bodies, as geological objects or forest features, etc., and their limits may be poorly constrained. Correspondingly, the environmental requirements concerning wetlands may be different or even inverse.

There is also much uncertainty about the boundaries of other water bodies (rivers, aquatic buffers, flooding zones, etc.), which impedes regulation of economic activity and conservation measures. For instance, it is unclear whether a lateral channel should be considered a part of the river at a valley width of 5 to 20 km. If yes, the questions are how to ensure the required environmental engineering for residential buildings and industrial facilities, and how to implement the rule prohibiting fuel storage and disposal of wastes. Other uncertainties concern approaches to water supply charging.

4 Facing Water Pollution Under Rapidly Changing Environment: Russia's Experiences

4.1 Run-Off Changes as a Water Quality Factor

Freshwater availability is the key quality criterion of water resources. Changes in water resources are discussed annually in the reports of the Ministry of Natural Resources of the Russian Federation (Government Report, 2019). Comprehensive scientific reviews can be found in (Ratkovich et al., 2003; Shiklomamov, 2008) for Russia as a whole and (Shiklomamov et al., 2013) for Siberia. These and other publications address patterns and features of water budget in large rivers and groundwaters within the Russian territory.

The changes have several causes: considerable reduction of freshwater supply and wastewater discharge in Russia in the 1990s after the breakdown of the Soviet Union; global warming of near-surface air; sea-level change in the Caspian Sea that affects a large populated territory. These factors might be correlated but the type of correlation is unclear, especially because some processes have opposite trends. The opposite trends are, for instance, that some valley glaciers are degrading while others remain quasi-stable, or groundwater levels may fall in wooded steppes (e.g., in the Altai offshoots) but rise in the taiga zone (e.g., 0.2–0.3 m since 1960s in West Siberia). Uncertainties concern, for example, the expansion of some lakes in northern taiga and tundra forests: whether it results from permafrost degradation or is a natural evolution trend of wetland ecosystems. Another example of uncertainty concerns level changes in the Caspian Sea for which predictions are poorly reliable (Shiklomanov et al., 2013; Savichev et al., 2018).

At the present state of knowledge, time shifts in seasons of high and low water level may be a reasonable explanation for the observed changes in water resources. Note that long-term annual means of atmospheric precipitation may remain invariable but more moisture can accumulate on watersheds in the season of lowest infiltration (late fall). Probably, large-scale correlated changes may occur in some lithosphere or hydrosphere zones. Evident man-caused changes to water resources occurred in the 2000s when the large Boguchany and Bureya hydroelectric power plants were put into operation. Other major power plants were constructed mainly in the 1950s through 1980s.

4.2 Source Control and Pollution Interception

For the past three decades, the structure of the economy in Russia has changed while the regulation of pollution became mandatory for water users. Meanwhile, point pollution sources still exist and discharge contaminants within the maximum level, though pollution dispersed by rain and meltwater from urban areas poses the highest risks.

The current practices of mitigating the risks are collecting the surface run-off in storm drains; channeling effluents to treatment systems; or setting up aquatic buffers along rivers and around lakes and restriction of economic activity within their limits. The run-off management requires changes to the urban infrastructure while creating the buffer zones may be problematic because the boundaries of water bodies remain poorly constrained and because most of the population lives near rivers and lakes. According to the Water Code (2018), the boundary of a river (bank line) is outlined according to the average level for the period of open water. At water level fluctuations of 8–10 m or more, many historic residential areas fall within the bank line. Moreover, progress of urban development leads to increasing costs of construction and operation of facilities for water treatment, sewage disposal, and removal of wastes and snow (in winter). The placement of wastewater treatment systems inside urban territories is regulated by sanitary legislation and incurs large expenses for creating buffers, while the problems of air, water, and soil pollution are still urgent. Nevertheless, contouring the aquatic buffers and upholding the related restrictions are subjects of yearly surveys and monitoring, which appears to be one of most efficient ways of pollution control.

Improvement of water supply, drainage, and canalization is crucial for urban life. The construction and operation of such systems is costly while accidents at water (and heat) supply facilities pose risks of flooding and impairment of sanitary conditions. In Russia the latter problem is aggravated by seasonal ground freezing (to 1 m or more) or the presence of permafrost and may worsen further upon climate warming.

In general, a special strategy in development of housing and public utilities may be required for Northern Eurasia. It appears reasonable to combine centralized and local systems of water and heat supply, to integrate park and buffer zones around water bodies, to use artificially rechargeable underground water reservoirs, etc. Improvement is required for the methods of wastewater treatment. In this respect, physical cleaning combined with biological cleaning and natural aeration was suggested (Savichev, 2010b) as an optimal strategy for the conditions of Siberia and northern European Russia (taiga and tundra forest landscapes). Management of controlled-drainage irrigation with underground confining is another important target for improvement.

4.3 *Biogenic Components in Sludge and Rehabilitation of Wetlands*

Wetlands cover a large part of the Russian Federation. They are quite poor ecosystems with suppressed vegetation that suffers from shortage of nutrients. On the other hand, some nutrient substances are present in wastewaters and in sludge remaining after treatment. Currently the users are charged for disposal of wastes, water drainage, and damage to peatbogs. However, peatlands in Siberia are expanding, and the peat thickness increases at 1 mm/yr, whereas the growth conditions of forests are worsening. In this respect, domestic wastewaters and specially prepared sludge, which appears to be an underestimated resource, may be used for adaptation of wetlands to the purposes of forestry. Of course, projects of this kind will require the respective technologies for transforming sludge into appropriate landfill material, but they will contribute to environmental quality over vast territories in Siberia. Improving forestation is important for climate and environmental sustainability. On the other hand, using sludge for farming or construction purposes may be risky and often poorly efficient.

4.4 River Flow Transfer

The idea of diverting a part of flow from Siberian rivers toward Central Asia appeared in the Soviet Union in the 1930s, and large-scale engineering systems for this transfer were designed and constructed in the 1930s through 1960s. Nowadays the idea is very unpopular, but it is fair to note that similar processes occurred also in nature, such as inputs of meltwater from valley glaciers into the Aral-Caspian basin. However, further development of this idea, and the feasibility of such projects, depends on global social and economic processes.

References

Alekin OA (1970) Fundamentals of hydrochemistry. Gidrometeoizdat, Leningrad, p 443. (in Russian)

Bogoslovsky BB (1960) Lake hydrology. Moscow State University, Moscow, p 335 (in Russian)

Bryzgalo VA, Nikanorov AM, Kosmenko LS, Reshetnyak OS (2015) Estuarine ecosystems of major rivers in Russia: human impact and environment. Southern Federal University, Rostov-on-Don, p 164 (in Russian)

Effluent Standards (2018) Standards of permissible discharges of substances and microorganisms into water bodies. Methodology. Approved by Decree No. 342 of 31.07.2018. Ministry of Nature Resources of Russia, Moscow, p 35 (in Russian)

Environment Law (2001) Federal law No 7-FL on environment conservation, adopted by the State Duma on 20.12.2001 and approved by the federation council on 26.12.2001. Ministry of Justice of the Russian Federation, Rossiiskaya Gazeta, p 6, no 2874. (in Russian)

Environment Law (2017) Federal law no 7-FL on environment conservation, as amended on 07/29/2017. Kremlin, Moscow, p 77. (in Russian)

Government Report (2019) Water resources in Russian federation: state and use in 2018. NIA-Priroda, Moscow, p 290. (in Russian)

Gusev SI (2019) Water cadastre. Surface water and groundwater resources: use and quality, 2018. Roshydromet, St. Petersburg, p 153. (in Russian)

Guseva NV, Savichev OG (2016) Hydrochemical balance of Itkul-Shira lake system (Khakassia, Russian Federation). Contemp Probl Ecol 9(5):590–599. https://doi.org/10.1134/S1995425516050061

Guseva TV (2007) Hydrochemical indicators of the environment: reference materials. FORUM-INFRA-M, Moscow, p 192 (in Russian)
Guseva NV (2018) Water chemistry controls in lakes and rivers of different landscapes and climate zones in fold areas of central Eurasia. Author's Abstract, PhD Thesis, Tomsk Polytechnical University, Tomsk, p 43. (in Russian)
Hendriks Martin R (2010) Introduction to physical hydrology. Oxford University Press, Oxford, New York, p 331
Karpukhin VF (2001) Handbook of modern technologies for the treatment of natural and waste water and equipment. Ministry of Nature Resources of Russia, Moscow, p 253. (in Russian)
Lisitsyn AP (1994) A marginal filter of the oceans. Oceanology 34(5):735–747
Liss OL, Abramova LI, Avetov NA, Berezina NA, Inisheva LI, Kurnishnikova TV, Sluka ZA, Tolpysheva TY, Shvedchikova NK (2001) Bogs of western Siberia and their conservation value. Grif i K, Tula, p 584. (in Russian)
Maloletko AM (2008) Mesozoic and cenozoic evolution of river systems in West Siberia. Tomsk State University, Tomsk, p 288. (in Russian)
Mikhaylov VN, Dobrovol'sky AD, Dobrolubov SA (2005) Hydrology (Gidrologiya). Vysshaya shkola, Moscow, p 463. (in Russian)
Pologova NN, Lapshina ED (2002) Carbon accumulation in peat bog deposits of the Vasyugan Swamp. In: The Vasyugan swamp. Current state and development. Institute of Atmospheric Optics SB RAS, Tomsk, pp 174–179 (in Russian)
Ratkovich DYa (2003) Problems of water supply. Nauka, Moscow, p 352.
Reymers NF (1990) Natural management: reference dictionary. Mysl, Moscow, p 637. (in Russian)
Sanitary Norms and Regulations (2000) Working document SanPiN 2.1.5.980–00. Sanitary rules and regulations. Hygienic requirements to surface water safety. Minzdrav RF, Moscow, p 23. (in Russian)
Sanitary Norms and Regulations (2001) Working document SanPiN 2.1.4.1074–01. Sanitary and epidemiological rules and regulations. Drinking water. Hygienic requirements to water quality of centralized water supply system. Quality Control. Minzdrav RF, Moscow, p 16. (in Russian)
Sanitary Norms and Regulations (2002) Working document SanPiN 2.1.4.1110–02. Sanitary rules and regulations. Sanitary buffer zones of water supply sources and water lines for drinking purposes. Drinking water and water supply of populated areas. Effective Date 01.06.2002 . Rospotrebnadzor, Moscow, p 12. (in Russian)
Savichev OG (2008) Channel deformation in the Vasyugan and Parabel catchments (West Siberia). Bull Tomsk Polytech Univ 313(1):82–87
Savichev OG, Paromov SV (2014) Swamping in the taiga zone of West Siberia: hydrological aspects. Bull Tomsk Polytech Univ 324(1):154–161
Savichev OG, Kolesnichenko LG, Saifulina EV (2011) The ecological-geochemical state of water bodies in the Taz-Yenisei interfluve. Geogr Nat Resour 32(4):333–336. https://doi.org/10.1134/S1875372811040056
Savichev OG, Mazurov AK, Pipko II, Sergienko VI, Semiletov IP (2016) Spatial patterns of the evolution of the chemical composition and discharge of river water in the Ob river basin. Dokl Earth Sci 466(1):59–63. https://doi.org/10.1134/S1028334X1601014
Savichev OG, Lgotin VA, Moiseeva YuA (2018) Conditions of groundwater level variations in upper hydrodynamic zone in Tomsk region. Geoecology. Geological Engineering. Hydrogeology. Geocryology 6:31–40. https://doi.org/10.1134/S0869780318060054
Savichev O, Soldatova E, Rudmin M, Mazurov A (2020) Geochemical barriers in oligotrophic peat bog (Western Siberia). Appl Geochem 113(104519):1–11. https://doi.org/10.1016/j.apgeochem.2019.104519
Savichev OG, Guseva NV (2020) Methodology of management of river basins geochemical balance in Western Siberia. Geo Assets Engineering Bull Tomsk Polytech Univ 331 (5) (in press)
Savichev OG (2010a) Water resources of the Tomsk region. Tomsk Polytechnical University, p Tomsk, 248. (in Russian)
Savichev OG (2010b) Discharge regulation in surface water bodies. Water: Chem Ecol 9:35–39

Savichev OG (2012) Theoretical foundations of environment conservation: students manual. Tomsk State University, Tomsk, p 126. (in Russian)
Shiklomanov AI, Lammers RB, Lettenmaier DR, Polischuk Y, Savichev OG, Smith LC, Chernokulsky AV (2013) Hydrological changes: historical analysis, contemporary status, and future projections. In: Groisman PYa, Gutman G (eds) Environmental changes in Siberia: regional changes and their global consequences, Chap. IV. Springer, Berlin, pp 111–154. https://doi.org/10.1007/978-94-007-4569-8_4.
Shiklomanov IA (2008) Water resources of Russia and their use. State Hydrological Institute, St. Petersburg, p 600. (in Russian)
Shvartsev SL (2008) Geochemistry of fresh groundwater in the main landscape zones of the earth. Geochem Int 46(13):1285–1398
State Standard (2001) Working document GOST 17.1.1.01–77. Environment safety. Hydrosphere. Water use and conservation. Basic terms and definitions. Standart, Moscow, p 15. (in Russian)
State Standard (2003) Working document GOST 27065–86. Water quality. Terms and definitions. Standart, Moscow, p 9. (in Russian)
Subsoil Use 2011. Federal law No 27-FL on Subsoil Use of 03.03.1995 as amended on 06.12.2011. Official Internet Portal of Legal Information www.pravo.gov.ru, 06.12.2011, 64 p (in Russian)
Wastes Management (2012) Federal law No 89-FL on management of production and consumer wastes adopted by the state duma 22.05.1998 and approved by the federation council 10.07.1998 as amended on 29.06.2012. Official Internet Portal of Legal Information www.pravo.gov.ru, 01.07.2012, p 20. (in Russian)
Wastewater Treatment (2015a) Information and technical handbook on the best available technologies. Wastewater treatment using centralized water disposal systems for settlements, Urban districts. ITS 10–2015. Rosstandart, Moscow, p 359. (in Russian)
Wastewater Treatment (2015b) Information and technical handbook on the best available technologies. Wastewater treatment in manufacture of products (Goods), Performance of works and provision of services in large enterprises. ITS 8–2015. Rosstandart, Moscow, p 77. (in Russian)
Wastewater Treatment (2016) Information and technical handbook on the best available technologies. Recycling and disposal of wastes (Except for Thermal Disposal (Incineration) of Wastes. ITS 15–2016. Rosstandart, Moscow, p 139. (in Russian)
Water Code of the Russian Federation (2018) Document No. 74-FL of June 3, 2004: as amended on 08/03/2018. Kremlin, Moscow, p 48. (in Russian)
Water Pollution (2002) Working document 52.24.643–2002. Methodoloical guidelines for estimating water pollution from water chemistry. Rosgidromet, Moscow, p 34 (in Russian)
Water Treatment Regulations (2012a) Working document SP 30.13330.2012. Water treatment rules and regulations. Domestic Water supply and drainage systems in buildings. Revised Edition SNiP 2.04.01–85*. Minregion of Russia, Moscow, p 57. (in Russian)
Water Treatment Regulations (2012b) Working document SP 32.13330.2012. Water treatment rules and regulations. Sewage. Pipelines and wastewater treatment plants. Revised Edition SNiP 2.04.03–85. Minregion of Russia, Moscow, p 82. (in Russian)
Zektser IS (2007) World groundwater: resources, use, forecasts. Nauka, Moscow, p 438. (in Russian)

Critical Assessment of Groundwater Overflow Zone for Sustainable Management in North East India

Malabika Biswas Roy, Jayanta Debbarma, Sudipa Halder, Pankaj Kumar Roy, Supriya Pal, Hirok Chaudhuri, and Mrinal Kanti Mondal

1 Introduction

Groundwater is one of the valuable sources to sustain human life on earth. Population explosion and industrial outgrowth have led to the over exploitation of water which will bring about massive changes in our future water supply (Alvis et al., 2019). But in some regions groundwater flooding is one of the serious concerns. Sometimes the ground flooding is misunderstood and overlooked which can be a biggest threat to the society (Huges et al., 2011). Ground flooding is mainly caused by the emergence of water from sub-surface permeable strata occurring under confined condition and is often characterized by long-term flood durations that can cause prolonged disruption and relatively high damage (Marsh et al., 2002). Such floods may also cause by leaking of water from nearby fault line or springs or certain diffuse

M. B. Roy
Women's College, Calcutta, Kolkata 700003, India

J. Debbarma
PWD (Water Resources), Agartala, Tripura 799 006, India

S. Halder · P. K. Roy (✉)
School of Water Resources Engineering, Jadavpur University, Kolkata 700 032, India
e-mail: pk1roy2014@gmail.com

S. Pal
Department of Civil Engineering, National Institute of Technology Durgapur, Durgapur, India
e-mail: supriya.pal@ce.nitdgp.ac.in

H. Chaudhuri
Department of Physics, National Institute of Technology Durgapur, Durgapur, India
e-mail: hirok.chaudhuri@phy.nitdgp.ac.in

M. K. Mondal
Department of Chemical Engineering, National Institute of Technology Durgapur, Durgapur, India
e-mail: mrinalkanti.mandal@che.nitdgp.ac.in

Y. Li et al. (eds.), *BRICS Countries: Sustainable Water Resource Management and Pollution Control*, https://doi.org/10.1007/978-981-99-9581-3_3

location emerging from sub-surface saturated permeable zones (Morris et al., 2012). Groundwater flooding occurs due to the extensive saturation of the subsurface layers caused by prolonged rainfall events. According to Abboud et al., 2018, Groundwater flooding is the result of water table rises when recharge exceeds the aquifer abstraction having low storativity character or may be due to the rising river stages into permeable, river-connected fluvial-alluvial aquifers. The occurrence of such flood events continues to cause damage and disruption (Kjeldsen et al., 2016). The loss of life and property is rare but it can have local and regional impacts on economic and social behavior of a particular region.

Few or no studies are there related to groundwater flooding at a global level. In Tripura, India it is estimated that groundwater overflow not only causes flooding or inundation of the area but also causes seasonal water scarcity and alteration in water quality. Huge amount of groundwater overflow occurs after longer periods of sustained high-intensity rainfall. So, higher rainfall leads more water to infiltrate into the ground causing the water table to rise above the original level. Generally, in Tripura, groundwater tends to flow from hills to the plains as a result of change in slope where the ground level fluctuates seasonally at a higher rate. In the low-lying plain, the water table is usually at a shallow depth, but during rainy periods, the additional groundwater flowing downwards from the uplands can cause the water table to rise to the ground surface causing extensive groundwater flooding (Macdonald, 2003). Groundwater overflow resulting in floods is most likely to occur in low-lying areas underlain by permeable rock. These are mainly extensive and regional aquifers, such as sandstone, or may be localized sands or river gravels in valley bottoms underlain by less permeable rocks. Groundwater flooding takes longer time to dissipate because the movement of groundwater is much slower than surface water and will take time to infiltrate and saturate the underground layers (Bingham, 2015).

Groundwater exploration can be made through quantitative assessment of geoelectrical data through resistivity method (Adagundo et al., 2018; Araffa et al., 2019). Past literature studies revealed that the Vertical Electrical Sounding (VES) has the potential to estimate the aquifer parameters which provide reasonable accurate results that can be used to understand the subsurface layers through groundwater exploration and will help to identify the groundwater potential zones (Sikander and Christen, 2012) and that must be verified with hydro-geological data (Debbarma et al.,2016a, 2016b and 2016c). It also stated that 2D interpretation of 1D VES measurements can produce improved subsurface geophysical images and represents a potential useful tool for larger scale geological investigations especially in the case of reprocessing existing VES data sets (Atzenoglon et al., 2012). Literature studies also stated that the litholog obtained from borehole and the VES output should not be treated as exactly the same but can save on cost and manpower (Roy et al., 2017) as quicker results can be obtained through the VES study (Felix et al., 2013; Debbarma et al., 2016a,2016b and 2016c). Therefore, the objective of the study is (a) to identify the groundwater overflow zones in the plain land and hilly areas and (b) to assess the impact of hydrological parameters on aquifer discharge for sustainable conservation.

2 Study Area Setting

Tripura is one of the states of India. It is located in the northeastern part of the Indian subcontinent. It is bounded by Bangladesh in the north, west, and south and Mizoram to the east and to the northeast by the state of Assam. It is one of the smallest of Indian states and is located in an isolated hilly region of the country comprising a number of indigenous tribes. Agartala is the capital of Tripura comprising an Area of 10,486 square km and Population of 3,671,032 according to 2011 census. Tripura area is now facing high demand in water supply than any other parts in the state due to high population density (Debbarma et al., 2016b, c, a). In Tripura, there are many rivers and most of them are almost dried up due to the higher amount of silt which they carry from higher altitudes and deposits downstream. Seven districts with sixty two Gram Panchayats (GPs) have been considered here to assess the overflow zone of Tripura and are given in a tabular format. Within the districts there are seven valleys viz., Udaipur Sabrum valley, Agartala valley, Khowai valley, Kamalpur valley, Kailasahar valley, Dharamnagar valley, and Amarpur valley (Fig. 1). It is found that most people are facing acute scarcity of water in few areas during non-monsoonal seasons in terms of both ground and surface water (Debbarma et al., 2013). Whereas in the monsoon season most of these regions are flooded through extensive groundwater overflow through rainfall events. Therefore, these regions can be identified and sustainable management of groundwater can be done through the implementation of recharge harvesting techniques. Thus it is needed to identify the overflow zone of flowing wells for assessing groundwater potential in Tripura in order to establish the optimum route of valleys using structural measures to avoid surface flooding or to minimize water quality deterioration to some extent with some suitable adaptation strategies.

3 Materials and Methods

In order to assess the groundwater overflow zone meteorological data such as rainfall evaporation, and evapotranspiration has been collected for the last twenty years (1996–2015) from the Indian Meteorological Department (IMD) and the historical discharge data of the major rivers has been collected from Central Water Commission (CWC), Government of India. For hydrogeological investigation, data like groundwater level, lithologs of different locations, specific yield and storativity were collected from Central Groundwater Board (CGWB). Original slopes of the land have been calculated from topographical maps which were collected from the Survey of India. Geological map of the study area was downloaded from the Geological Survey of India (GSI) website. Thematic maps were prepared with the help of AutoCAD software. Few lithologs data was collected from Public Works Department (PWD) and Drinking Water & Sanitation, Govt. of Tripura and verified accordingly with the help of actual field survey through Vertical Electrical Sounding. The VES was carried out at different locations (Fig. 2) and analysis was done using Schlumberger

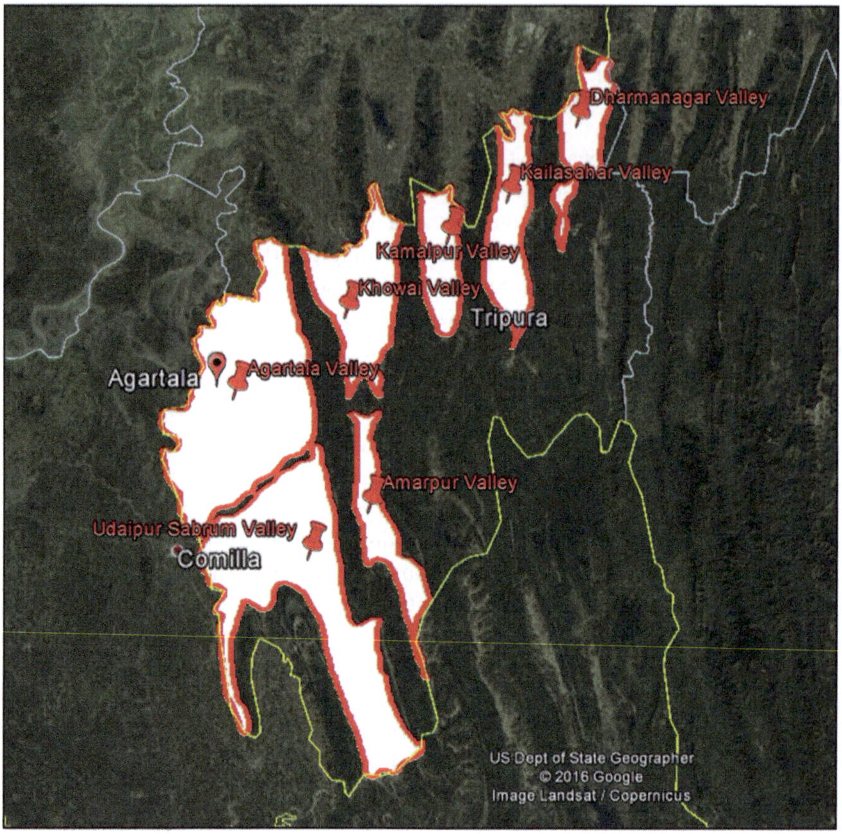

Fig. 1 Location of the study area and delineated districts of Tripura with seven valley

technique. Deep tubewell borings at different locations and depths were carried out by using the rig (reverse/direct) machine. Pumping tests have been conducted for the continuous operation of 48 h and accordingly drawdowns are observed with respect to time to estimate the aquifer parameters using different formulae (Roy et al., 2015). The methodology is represented in the flowchart below (Figs. 1 and 3).

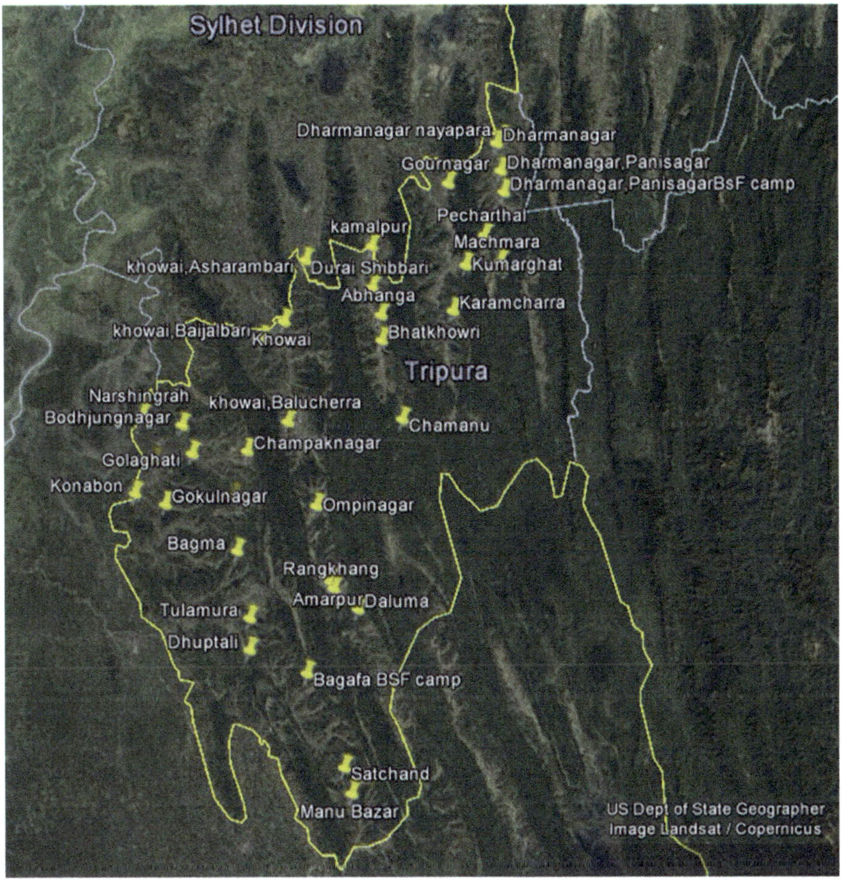

Fig. 2 Locational point for the electrical resistivity survey

4 Results and Discussion

4.1 *Valley-Wise Geological Characteristics and Identification of Water Layer Zone*

The litholog sites are plotted and their latitudinal and longitudinal extent has been marked using Google earth (Fig. 2). Valleys are delineated using topographical map with the ArcGIS software version 10.3. Lands are classified based on the deep tube well, marshy land, and groundwater level to identify the overflow zone of Tripura highlighted in Figs. 4 and 5. Based on the litholog and VES survey (Figs. 5, 6, 7, 8, 9, 10, 11, 12, 13, 14 and 15) and using AutoCAD, the geological cross-sectional fence diagram of various locations of the valley has been prepared and it has been found that the water layers are differentiated along the distance and it runs parallel through

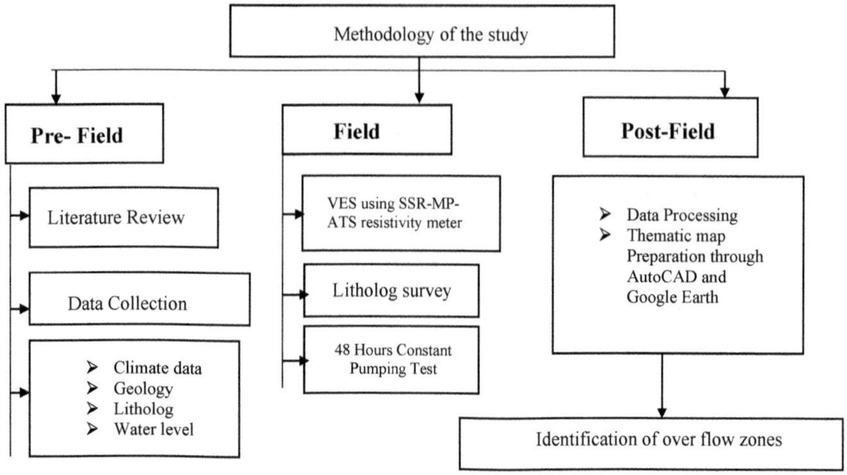

Fig. 3 Flowchart of the methodology

the ground surface for almost all valleys. The figures also depict that every thin layer is connected with thick layers that represent all water layers which are interconnected and groundwater flow moves upward to downward along with river slope direction which is parallel to surface water depth and causes flood during extreme rainfall events. In Tripura all the rivers are rain-fed and no ice melt water is available in the state and two or three major waterfalls can be found and mini periodical waterfalls are available which rejuvenate during rainy season at top of the hills. The figures are also highlighting that piezometric head mainly increases during rainy season and decreases during non-rainy periods. At the plains land of the valley referred in Fig. 5, overflow water is available throughout the year which usually comes from static groundwater level (<50 m). The dynamic groundwater level increases to about 2 to 3 m in plain land during rainy seasons whereas about 20 to 100 m decrease in the water level can be seen in the hilly areas and this shows that there is a strong relationship developed between static water layer and dynamic water layer due to downward movement (hilly zone to flat zone) where gradient also plays an important role.

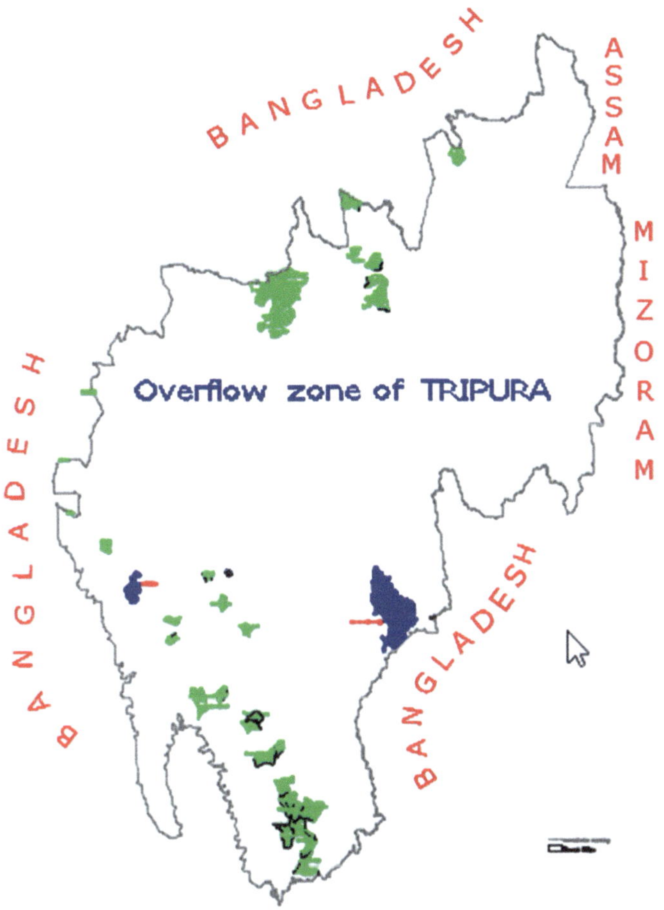

Fig. 4 Overflow zone of Tripura

4.2 *Estimated Overflow Zones of Tripura*

Based on the extensive groundwater exploration overflow zones of Tripura have been demarcated and groundwater discharge of different sites has been calculated. The average aquifer parameters of the Tripura state based on the pumping test have been given in a tabular format (Table 1). The litholog character of the different sites has been shown in Fig. 16. The artesian head ranged between 0.3 and 1.5 m (Table 2) which falls under green zone (Fyfe et al., 2005). Therefore, the pipe would need to be extended more than the present artesian head above the ground level. In general, a 30 percent bentonite grout may be used for flowing well construction or repair. A general overview of flowing artesian wells has also been shown in Fig. 16 and is prepared to assess the discharging groundwater above the top of the well casing having the variation in geological and topographical features. The discharge is found

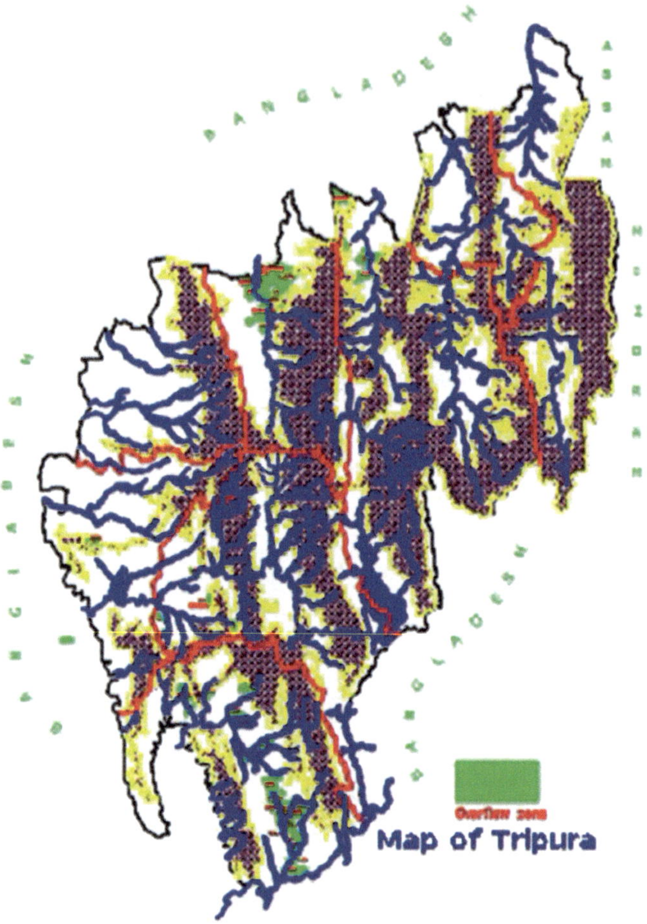

Fig. 5 Overflow zone of Tripura along with flat portion of the valley

to be 29.5 m^3/h considered as lower with an aquifer thickness of 28.5 m whereas the maximum discharge is observed as 136.3 m^3/h having almost the same thickness reflected in Table 2. Therefore, valley-wise variation in the potential zone and it was found that by having the same aquifer thickness, few are required to lower the pipe. Therefore understanding the aquifer parameters is most essential and the regions having the more hydraulic conductivity would be the recharge and greater will be the discharge during high rainfall events. The regions where the discharge is higher are the areas where groundwater flooding is predominant and sustainable management techniques are the most required.

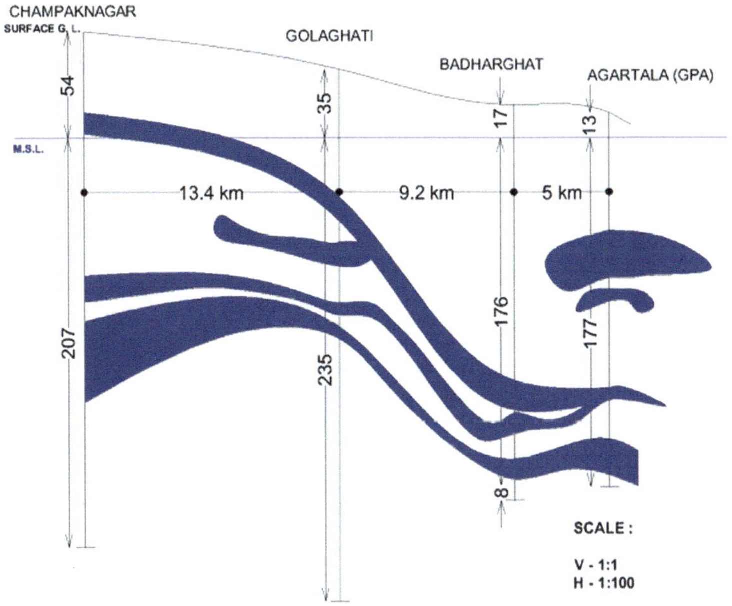

Fig. 6 Water zone of Dharmanagar valley

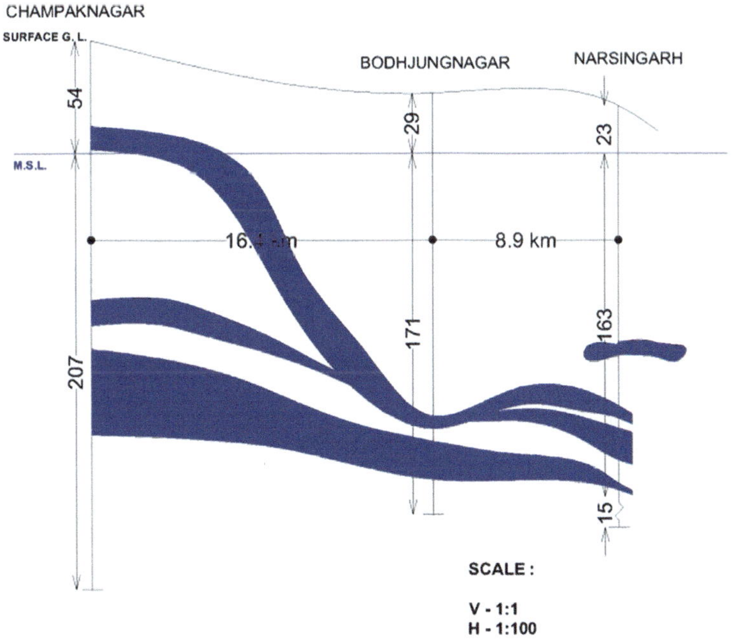

Fig. 7 Water zone of Kailasahar valley

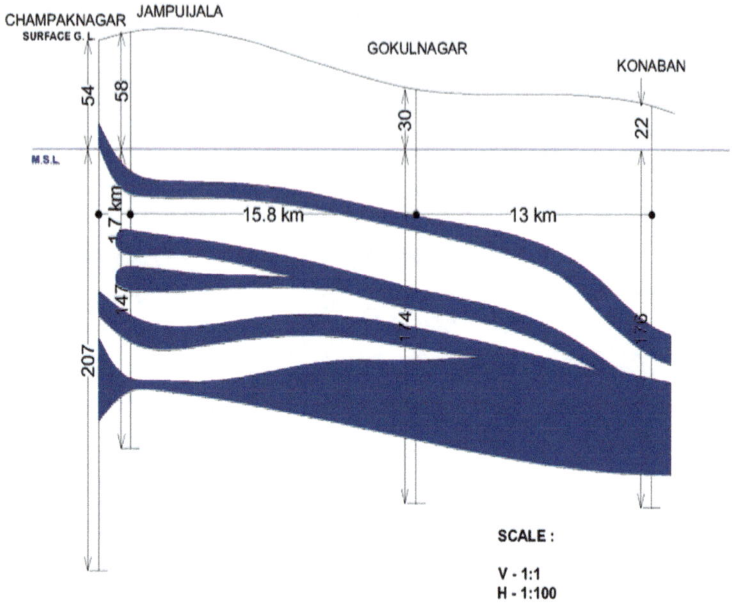

Fig. 8 Water zone of Kamalpur valley

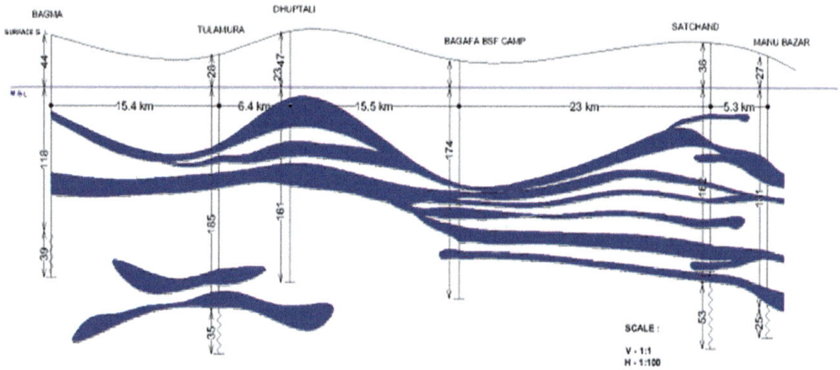

Fig. 9 Water zone of Khowai valley

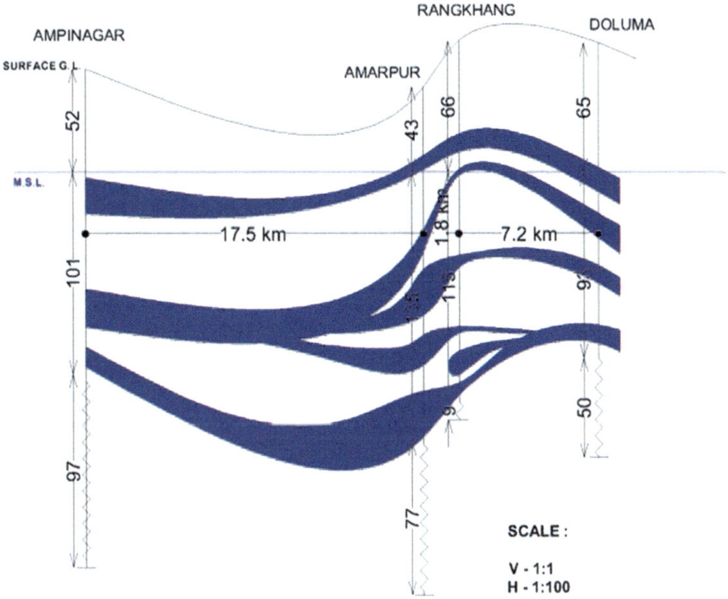

Fig. 10 Water zone of Udaipur valley

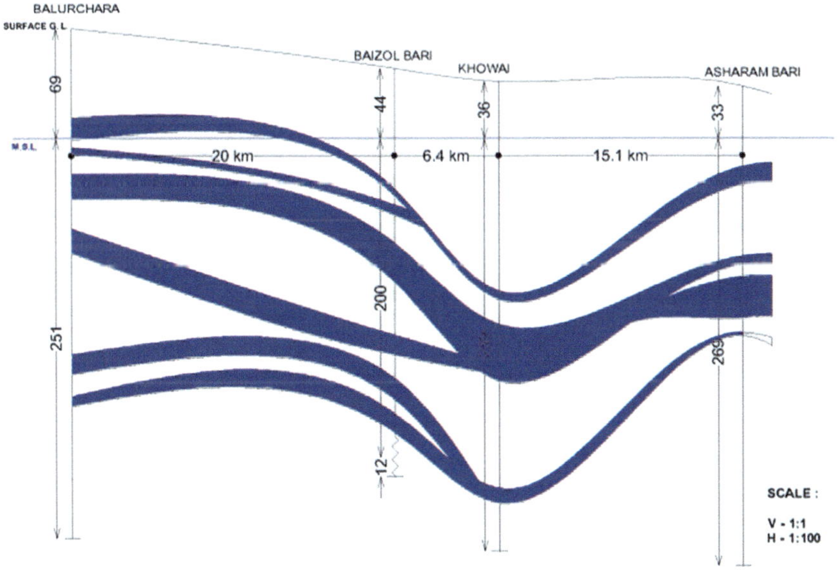

Fig. 11 Water zone of Amarpur valley

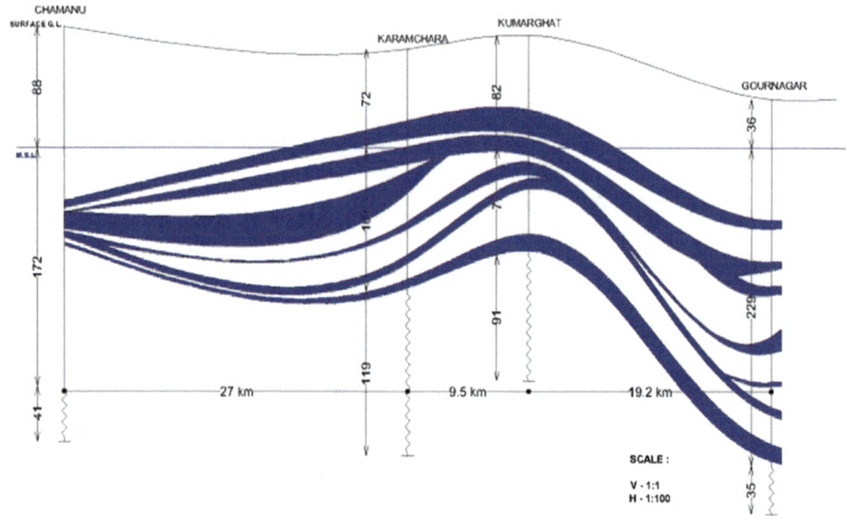

Fig. 12 Water zone of Agartala valley 1 Champaknagar Gokulnagar

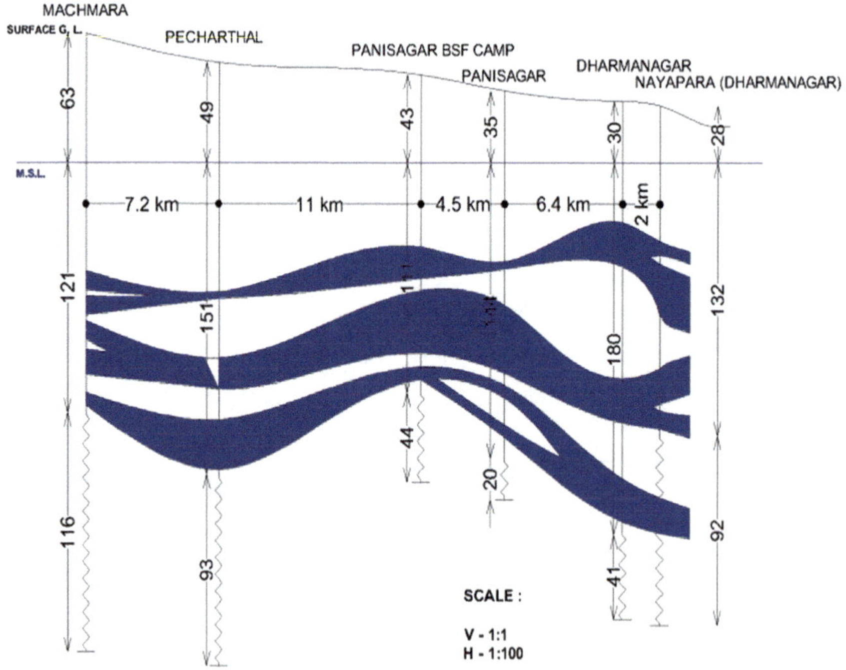

Fig. 13 Water zone of Agartala valley 2 Champaknagar Narsingarh

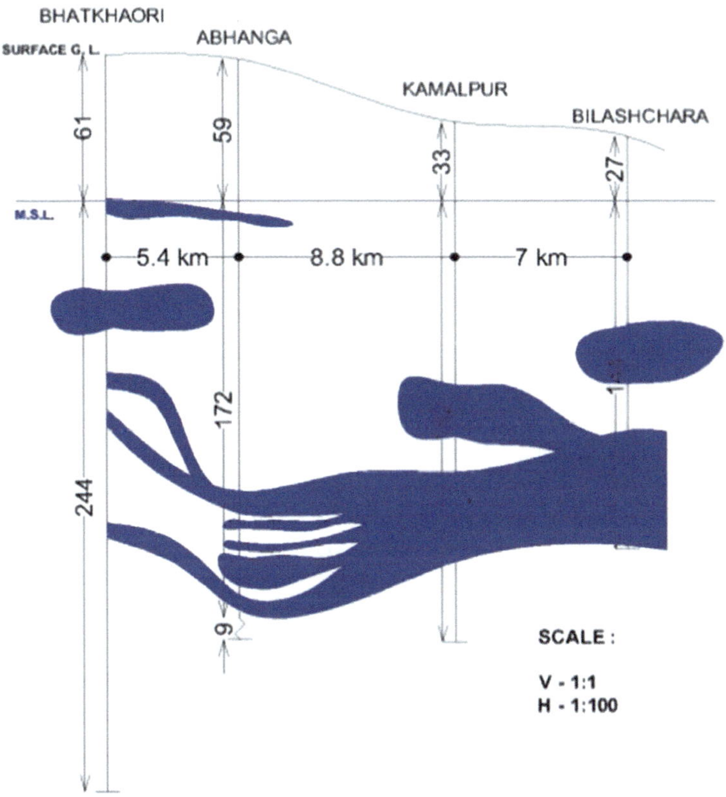

Fig. 14 Water zone of Agartala valley 1 Champaknagar Agartala

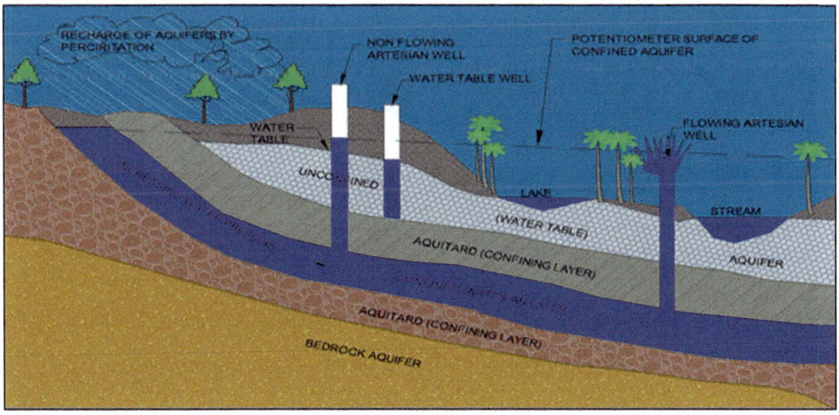

Fig. 15 Typical overflow zone of Tripura state prepared using AutoCAD

Table 1 Average aquifer parameters of Tripura state as estimated from pumping test data (Debbarma et al., 2016a, 2016b, 2016c)

State	Pumping rate (m^3/hour)	Storativity (s)	Transmissivity (T) m^2/day	Hydraulic conductivity m/day
Tripura	80	0.00027–0.00059	1100–1689	15–28

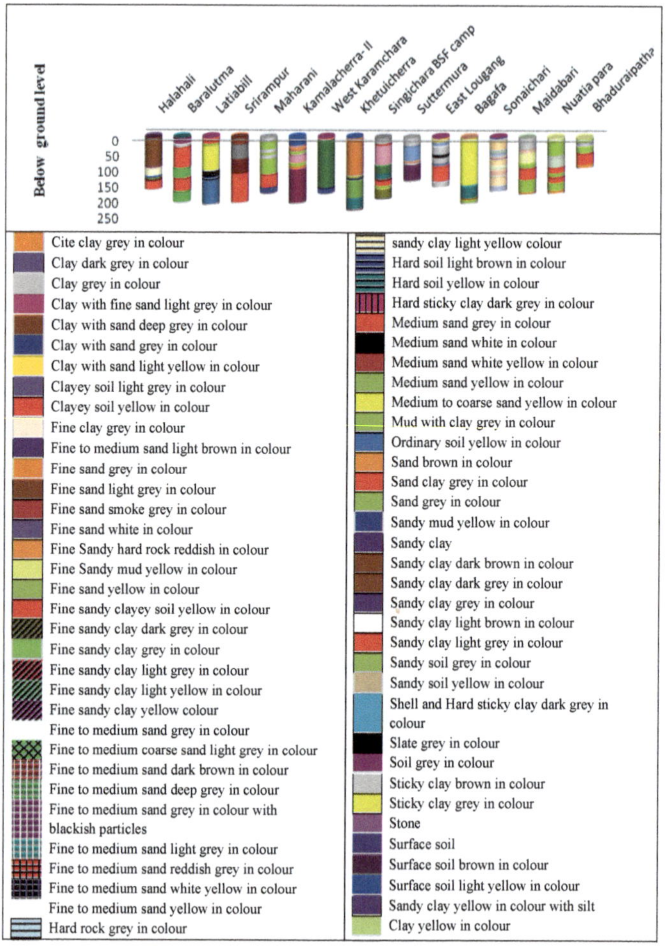

Fig. 16 Lithological strata of overflow borehole at different locations in Tripura, (Debbarma, 2018)

Table 2 Groundwater information of overflow borehole of Tripura

Sl. No	GP/VC	Name of valley	Depth of borehole (m- BGL)	Depth of pipe lowering (m^{-1}BGL)	Filter layer depth (m-BGL)	Discharge (m^3/h) (thickness of aquifer in m)	[b]Artesian head(m^{-2}AGL)
1	Halahali	Kamalpur	170.7	201.2	102.1–139	68.2 (36.9)	0.5
2	Baralutma	Kamalpur	213.4	136.5	104.8–139.4	29.5 (28.5)	0.6
3	[a]Latiabill	Kamalpur	219.5	140.8	110.4–137.7	68.2 (20.2)	0.8
4	Srirampur	Kamalpur	213.4	205.2	167.1–202.1	68.2 (35.0)	0.7
5	Maharani	Kamalpur	182.9	164.8	125.2–161.7	81.8 (36.5)	1.0
6	[a]Kamalacherra II	Kamalpur	213.4	106.7	54.9–103.9	59.1 (32.6)	0.4
7	West Karamcherra	Kailasahar	183	165.5	125.9–162.4	90.9 (36.6)	0.5
8	Khetuicherra	Kailasahar	237.8	231.4	198.6–228.3	22.8 (29.7)	1.3
9	Singicherra BSF Camp	Khowai	198.1	158.0	121–157.4	136.3(36.4)	1.5
10	Sutarmura	Agartala	152.4	138.8	99.2–135.7	68.2 (36.5)	0.7
11	East Lowgang	Udaipur Sabrum	158.5	146.4	106.7–143.2	68.2 (36.5)	0.6
12	Bagafa	Udaipur Sabrum	213.4	202.6	163–199.5	68.2 (36.5)	0.7
13	[a]Sonaicherra	Udaipur Sabrum	173.8	159.2	110.4–156.5	52.2 (36.8)	0.3
14	Maidabari	Udaipur Sabrum	182.9	140.2	100.7–137.5	54.0 (36.5)	0.6

(continued)

Table 2 (continued)

Sl. No	GP/VC	Name of valley	Depth of borehole (m- BGL)	Depth of pipe lowering (m-[1]BGL)	Filter layer depth (m-BGL)	Discharge (m^3/h) (thickness of aquifer in m)	[b]Artesian head(m-[2]AGL)
15	Noatia para	Udaipur Sabrum	182.9	152.5	100.6–118.9	59.5 (36.7)	0.4
16	Bhaduraipathar-w/s	Amarpur	97.5	92.0	52.1–89.0	56.8 (36.9)	0.4

[a] *Filter layer depth is not continuous;*
[b] *Indicates hydraulic water pressure developed through pipe;*
1: *Below Ground Level;* 2: *Above Ground Level*

5 Conclusion

In this study, the groundwater overflow zones have been demarcated using hydrogeological explorations in the Tripura state of India using secondary data and primary data as collected from the field. Groundwater is unequally distributed throughout the state and therefore requires proper management. Overflow zone called as "flowing artesian well" has been confined by impermeable materials and the static water level is found to be on top of the aquifer and the ground-level surface. The results showed that overflow zones are found maximum in flat portion compared to hilly areas because the movement of water is from upward to downward. The higher amount of uncontrolled discharge with good water pressure and limited aquifer thickness in some valleys having flowing wells may cause flooding of the nearby areas and also aquifer materials may come along with the groundwater and eventually may reach surface water bodies and thus creating alternation in the quality of surface water. This paper also concludes that the uncontrolled overflow zones from wells may cause water scarcity during low rainfall periods if not properly utilized in a sustainable manner.

6 Recommendation

- Micro-water storages in the hilly areas could be made by combining two valleys which will not only enhance the storage capacity but will also increase recharge volume.
- Since the movement of water comes from upstream to downstream so water withdrawal should have to be limited particularly in hilly terrain to avoid landslide and will also increase the overburden pressure in the overflow zones.
- If the withdrawal rate is found to be higher, the soil materials will be dislodged faster with more granular space between the soil particles with following elastic and geometric property resulting secondary settlement that may occur due to heavy loads. Such condition is needed to be managed.
- It is also observed that almost all flat areas are found to be overflow zone so more water could be extracted using extra pumping arrangement and to be stored in a separate reservoir for future use in water scarcity areas
- Since all streams present in Tripura are almost perennial to some extent, still Tripura has a water scarcity in some patches which might be due to the fluctuation level of water table varying in a wider range seasonally. So to minimize these problems more water is needed to be stored and construction of surface water retaining structures could be an alternative measure in or nearby artesian aquifer zone to mitigate the water crisis for a longer period.

- Gate valve may be installed at all uncontrolled discharge of deep tube well which are supposed to be in the overflow zone in order to mitigate the water crisis for scarcity areas. The present study will also incur some valuable techniques to manage the groundwater potential zone to a sustainable extent.

References

Abboud JM, Ryan MC, Osborn GD (2018) Groundwater flooding in a river-connected alluvial aquifer. J Flood Risk Manag 11:e12334

Adagunodo TA, Margaret KA, Lukman AS, Ahzegbobor PA, Kehinde DO, Felicia OA (2018) Groundwater exploration in Aaba residential area of Akure, Nigeria. Front Earth Sci 6:66. https://doi.org/10.3389/feart.2018.00066

Araffa SAS, Mahmoud IM, Hassan SS, Mustafa ST (2019) Geophysical interpretation for groundwater exploration around Hurghada area, Egypt. NRIAG J Astron Geophys 8(1):171–179. https://doi.org/10.1080/20909977.2019.1647389

Atzemoglou A, Tsourlos P (2012) 2D Interpretation of vertical electrical soundings: application to the Sarantaporon Basin (Thessaly, Greece). J Geophys Eng 9:50–59

Bingham A (2015) Groundwater and surface water flooding–statement, Land at Barnham, Eastergate & Westergate, BEW Landowner Consortium, A072162–1.

Debbarma J, Ghosh AK, Roy PK, Banerjee G (2016a) Study of identification of effective sand bed in aquifer zones using resistivity survey in Tripura: case studies. Am J Eng Res 5(1):111–120

Debbarma J, Roy PK, Pal M, Mazumdar A (2016b) Estimation of aquifer thickness of groundwater using resistivity survey in Tripura. Indian J Environ Prot 36(8):628–636

Debbarma J, Roy PK, Halder S, Banerjee G, Pal M (2016c) Estimating groundwater volumetric mass balance with hydraulic head using groundwater modeling system in Tripura, India. Asian J Curr Res 1(1):19–29

Debbarman J, Roy PK, Mazumdar A (2013) Assessment of dynamic groundwater potential of Agartala municipality area. Int J Emerg Trends Eng Dev 3(1):220–231

Debbarman J (2018) A study on groundwater assessment, Development and Management in the State of Tripura, Ph.D. Thesis awarded in Jadavpur University, Kolkata, India

Felix O, Oladimeji O, Ademilua L (2013) A correlation of Wenner array geo-electric log with borehole lithologic log. Res J Eng Appl Sci 2(2):129–130

Fyfe J, Martin D, Simpson M, Epp P, Carriou T (2005) Flowing artesian well-ministry of environment, British Columbia, Water Stewardship Information Series, ISBN 978–0–7726–7034–2

Hughes AG, Vounaki T, Peach D, Ireson A, Jackson C, Butler A, Bloomfield J, Finch JW, Wheater HS (2011) Flood risk from groundwater: examples from a chalk catchment in southern England. J Flood Risk Manag 4:143–155. https://doi.org/10.1111/j.1753-318X.2011.01095.x

Kjeldsen T, Prosdocimi I (2016) Assessing the element of surprise of record-breaking flood events. J Flood Risk Manag. https://doi.org/10.1111/jfr3.12260

López-Alvis J, Carrera-Hernández JJ, Levresse G et al (2019) Assessment of groundwater depletion caused by excessive extraction through groundwater flow modeling: the Celaya aquifer in central Mexico. Environ Earth Sci 78:482. https://doi.org/10.1007/s12665-019-8497-4

Macdonald D (2003) UK Groundwater Forum, Raising awareness groundwater. http://www.groundwateruk.org/FAQ_groundwater_flooding.aspx

Marsh TJ, Dale M (2002) The UK floods of 2000–2001: a hydro-meteorological appraisal. Water Environ J 16(3):180–188

Morris SE, Cobby D, Zaidman M, Fisher K (2012) Modelling and mapping groundwater flooding at the ground surface in Chalk catchments. J Flood Risk Manag. 11:S251–S268

Roy PK, Roy SS, Giri A, Banerjee G, Majumder A, Mazumdar A (2015) Study of impact on surface water and groundwater around flow fields due to changes in river stage using groundwater modeling system. A Clean Technol Environ Policy 17(1):145–154
Roy PK, Chaudhuri S, Halder S, Hossain S, Banerjee G, Debbarma J (2017) Developing a correlation for estimation of aquifer layer using resistivity survey with lithological logs in critical terrain condition. Asian J Water Environ Pollut 14(1):9–17
Sikandar P, Christen EW (2012) Geo-electrical sounding for the estimation of hydraulic conductivity of alluvial aquifers. Water Resour Manag 26(5):1201–1215

Performance Evaluation of Seismic Resisting Potential of Geo-Composite Liner in Waste Containment Structures–Some Case Studies

Supriya Pal, Avishek Adhikary, Hirok Chaudhuri, Mrinal Kanti Mandal, Kashyap Kumar Dubey, Pankaj Kumar Roy, and Malabika Biswas Roy

1 Introduction

Disposal of untreated or partially treated industrial and domestic wastes directly on land or any containment structures and landfill sites form leachate at the base of the deposited wastes, which is the primary concern of soil and groundwater pollution due to the migration of pollutant plume through sub-surface soil media and mixing with near-surface aquifers. Groundwater quality and lithospheric pollution level of surrounding sites are necessary to examine in a geo-environmental way, and the extent of pollution in the case of existing systems and for prevention of migration of pollutants by low-cost clay as liner materials also are to be evaluated judiciously. Secondly, side slopes failure of waste containment structures may occur during their construction, waste filling, or even post-closure of the systems. The failure of the containment structures can have catastrophic consequences, including

S. Pal · A. Adhikary
Department of Civil Engineering, National Institute of Technology Durgapur, Durgapur, India

H. Chaudhuri (✉)
Department of Physics, National Institute of Technology Durgapur, Durgapur, India
e-mail: hirok.chaudhuri@phy.nitdgp.ac.in

M. K. Mandal
Department of Chemical Engineering, National Institute of Technology Durgapur, Durgapur, India

K. K. Dubey
School of Biotechnology, Jawaharlal Nehru University, New Delhi, India

P. K. Roy
School of Water Resources Engineering, Jadavpur University, Kolkata, India

M. B. Roy
Women's College, Calcutta, Kolkata 700003, India

Y. Li et al. (eds.), *BRICS Countries: Sustainable Water Resource Management and Pollution Control*, https://doi.org/10.1007/978-981-99-9581-3_4

loss of life, property damage, and surface and groundwater contamination. Therefore, the stability of the containment structures must also be assessed judiciously to safeguard against unforeseen disasters.

2 Materials and Methods

The present studies aimed to examine contaminant transport through the landfill, along with the assessment of contaminant-soil interaction phenomena to minimize the migration of leachate. Laboratory studies were performed to explore the physical properties, composition, and leachate characteristics of wastes collected from the contaminated sites. The suitability of compacted geo-composite liner for containment of waste generated from waste containment structures was assessed by conducting various batch and column studies in the laboratory. The numerical modeling was also performed using the solute transport model HYDRUS 1D (Pal et al., 2014b) and finite element-based PLAXIS model to evaluate the thickness of the liner and safe side slope angle of the containment structures based on laboratory-derived test results as input parameters. The study will help develop any futuristic containment system for waste disposal containing toxic pollutants.

3 Results and Discussions

Case Study I:

The site was considered in proximity of an effluent discharge site (23,032′19 ″N and 87,014′49″ E) of a steel plant situated in West Bengal, India, which indicated significant lithospheric pollution due to the release of phenolic wastewater from the coke oven and subsequent migration to groundwater through subsurface geomedia (Ahmaruzzaman and Sharma, 2005) that poses a threat to human health and the surrounding environment (Mohan et al., 2005).

The percentage of soil fines (clay and silt) in the study area was 58 percent. The soil is graded as silt with low compressibility (ML) per the Indian Standard classification system (IS: 1498–1970). The soil was sparsely alkaline (pH 7.23). The studied soil cannot be considered a good candidate for barrier material for the retardation of wastewater migration due to its high saturated hydraulic conductivity ($K_s = 5.25*10^{-4}$ cm/s). The soil should possess low hydraulic conductivity (<1.0*10–7 cm/s) for considering as liner material in waste containment structures (Guler and Avci, 1993).

A laboratory model was developed for conducting column study experiments with solute as laboratory-prepared phenolic solution and adsorbent as different masses of soil columns (Ghosh et al., 2014; Mitra et al., 2015). The effluent solution was collected from the bottom of the column to observe the removal rates and also to generate breakthrough time. The outcomes of column tests are shown in Table 1.

Figure 1 shows the HYDRUS-1D predicted phenol breakthrough curves (BTCs) in soil columns for soil bed heights of 5, 10, and 15 cm. HYDRUS model projected phenol breakthrough using presumed tortuosity and dispersivity values. The predicted linear isothermal parameters closely matched the experimental BTCs for all the bed depths. The phenol BTC (Fig. 2) indicates that the point of exhaustion was achieved within 12 days of the period elapsed if this model were to be expanded to the current location of study where the groundwater table is 1.2 m below the ground surface. The study shows that in-situ soil has poor buffering resistance to retard the migration of phenolic pollutants from the wastewater outlet system to the groundwater. Therefore, it is necessary to replace the present studied soil with an adequately designed compacted clay liner to attenuate phenolic contaminants in the study area and protect precious groundwater against contamination.

Case Study II:

Table 1 Adsorptive process parameters in soil column with three different bed depths with 10 mg/L of influent phenol concentration

Z (cm)	t_b(h)	t_e(h)	q_{ad} (mg)	q_{total} (mg)	Removal (%)	Z_m (cm)
5	8	35	0.89	2.1	42.26	3.85
10	14	55	1.47	3.3	44.67	7.45
15	8	75	2.16	4.32	49.89	11.4

q_{total}: The total quantity of phenol migrating through the column; q_{ad}: Quantity of phenol adsorbed in the column; t_e: exhaustion time; t_b: initial breakthrough time; Z_m: length of the mass transfer zone for various bed depths (Z)

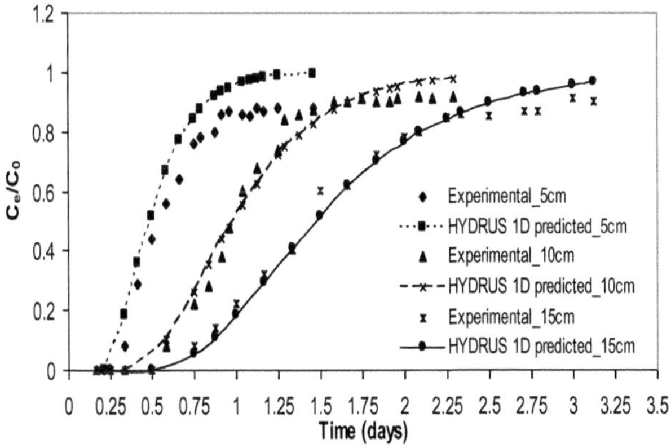

Fig. 1 Experimental and HYDRUS predicted plot of the aspect ratio of phenol against time in vertical column test

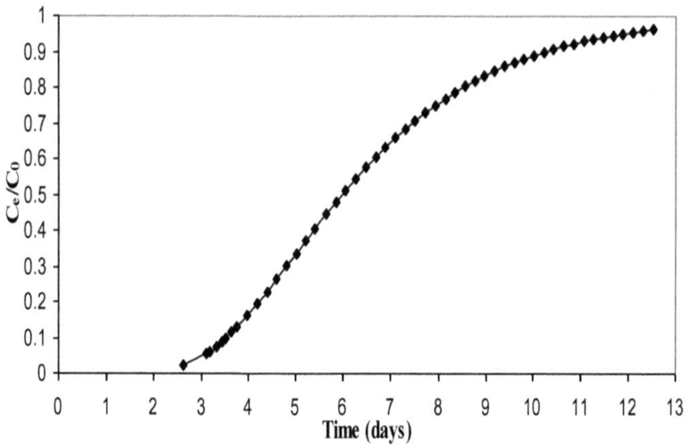

Fig. 2 HYDRUS predicted breakthrough curve for a soil bed of 1.2 m

In order to retard the sub-surface migration of heavy metal contaminants present in the leachate of ash pond sites, research was conducted to assess the performance of compacted laterite soil amended with bentonite and fly ash (Alam et al., 2012) used as composite liner material for ash pond sites of Thermal Power Plant. The ash was collected from three different ash pond sites in and around the industrial city of Durgapur, West Bengal, India of three thermal power plants: Mejia Thermal Power Station, Damodar Valley Corporation; NTPC—Sail Power Company Limited; Durgapur Projects Limited. The laterite soil samples were collected from Kamalpur, about 10 kms from Durgapur, West Bengal, India (Fig. 3).

The Cadmium concentrations found in the leachates were in the range of 0.127 to 0.37 ppm, whereas, according to IS 10500: 2012, the maximum permissible limit is 0.003 ppm, with no relaxation in the absence of alternate sources also.

The batch adsorption test was done as proposed by Mohammed and Naik, (2011), and heavy metal removal efficiency of the laterite soil was observed as 94% for Cadmium. In the case of the amended Soil (laterite soil and bentonite), the heavy metal removal efficiency was observed at 96% for Cadmium. So, a definite increase was observed.

Slope Stability Analysis

Using the HYDRUS software package—The pseudo-static slope stability analysis was performed using the SLOPE CLASSIC module of HYDRUS software. Both static and dynamic analyses were carried out. In the case of dynamic analysis, the earthquake effect is considered, with the help of vertical and horizontal seismic coefficients (Kh and Kv) as suggested elsewhere in the literature (Melo and Sharma, 2004). The analysis was performed using the four methods in the module viz. Bishop, Fellenius, Spencer, and Morgenstern-Price. The factor of safety values for two different dyke heights is shown in Table 2. The dyke height was raised to enhance the capacity of the existing ash pond.

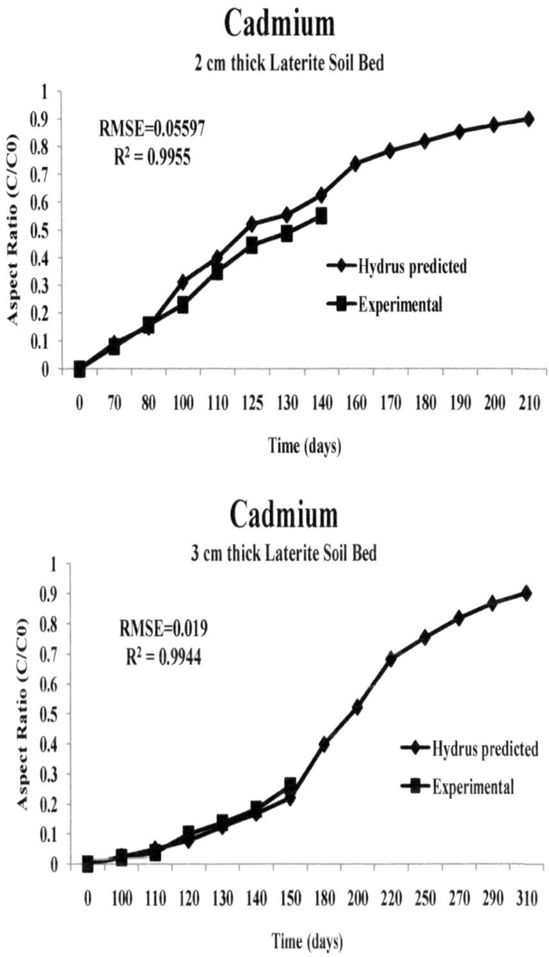

Fig. 3 Solute transport through laterite soil

Using PLAXIS software package—The stability of slopes is mainly assessed using the factor of safety. The strength reduction method was considered in PLAXIS software to ascertain the factor of safety. Geogrids are used as reinforcement materials to resist the elevated values of the strain and to increase the factor of safety. The dynamic analysis has been done by taking the acceleration time history of the real seismic record "Bhuj 2001, India" as an input motion (Table 3).

The factor of safety was reduced by 11.42% when the dump slope was raised by 6 m. A critical safety factor was obtained when the height of the embankment raised to 15 m, as the value is less than 1.5.

Table 2 Factor of safety values for the stability of slopes using different methods of analysis

Method	Bishop	Fellenius	Spencer	Morgenstern-price
Kh = Kv = 0	(a)2.48 (b)1.96	(a)2.43 (b)1.92	(a)4.08 (b)3.06	(a)4.01 (b)3.06
Kh = Kv = 0.1	(a)2.44 (b)1.88	(a)2.41 (b)1.84	(a)3.96 (b)2.85	(a)3.81 (b)2.83
Kh = Kv = 0.2	(a)2.18 (b)1.81	(a)2.18 (b)1.75	(a)3.45 (b)2.67	(a)3.21 (b)2.64
Kh = Kv = 0.3	(a)1.97 (b)1.70	(a)1.96 (b)1.65	(a)2.89 (b)2.46	(a)2.62 (b)2.45
Kh = Kv = 0.4	(a)1.64 (b)1.62	(a)1.16 (b)0.65	(a)2.04 (b)2.23	(a)2.26 (b)1.36

(a) Dyke Height = 6 m, (b) Dyke Height = 10 m

Table 3 Factor of safety of geogrid reinforced slopes with increasing dyke height

Height of embankment (slope angle = 17°)	8 m	10 m	12 m	14 m	15 m
Factor of safety	1.75	1.65	1.58	1.55	1.43

4 Conclusions

Though the laterite soil has good potential to be considered as liner material in the ash ponds due to its reasonably higher metal adsorption capacity and low hydraulic conductivity, further amendment of the soil using fly ash and bentonite is needed to lower its hydraulic conductivity and increase the shear strength. HYDRUS 3D finite-difference solute transport software's numerical modeling demonstrates the efficiency of modified soil over laterite soil as the recommended liner material.

The safety factor increases considerably by applying geogrids in the ash dyke. Hence, it can be concluded in the construction of steep and high slopes, reinforcing the embankment with geogrids can be useful. The slip surface shifts from the toe and becomes narrower when geogrids are incorporated, which reduces the chances of slope failure. Responses of the ash dyke under the dynamic input motion are observed, and displacement, acceleration, and velocity at the crest are observed to be more than the toe of the dyke. The ash dykes are more stable when the geogrids are installed at the toe level rather than the top (crest) level under dynamic loading. The slope angle can be increased from 17° to 35° by using geogrids at 2 m spacing, with the desired factor of safety of 1.5. The volume of the ash pond can be increased by up to 46.5% by applying the proposed condition.

References

Ahmaruzzaman M, Sharma DK (2005) Adsorption of phenols from wastewater. J Colloid Interface Sci 287(2005):14–24

Alam J, Khan MA, Alam MM, Ahmad A (2012) Seepage characteristics & geotechnical properties of fly ash mixed with Bentonite. IJSER 3(8)

Ghosh S, Mukherjee SN, Pal S, Mondal G (2014) Performance of a clayey soil used for landfill liner material in a tannery sludge disposal site for chromium attenuation. J Hazard Toxic Radioact Waste 18(4):04014023, 1–7.

Güler E, Avci C (1993) Comparison of Turkish and USA regulations in terms of landfill liner design (in Turkish). In: Proceedings of the sixth national clay symposium. Bogazici University, Istanbul, Turkey, pp 181–186

Melo C, Sharma S (2004) Seismic coefficients for pseudostatic slope analysis. In: 13th world conference on earthquake engineering Vancouver, B.C., Canada

Mitra S, Thakur LS, Rathore VK, Mondal P (2015) Removal of Pb(II) and Cr(VI) by laterite soil from synthetic wastewater: single and bi-component adsorption approach. Desalin Water Treat 2015:1–11

Mohammed SAS, Naik M (2011) Utilization of red soils and amended soils as a liner material for attenuation of copper from aqueous solution: isotherm and kinetic studies. J Environ Sci Technol 4(5):504–519

Mohan D, Singh KP, Singh VK (2005) Removal of hexavalent chromium from aqueous solution using low-cost activated carbons derived from agricultural waste materials and activated carbon fabric cloth. Ind Eng Chem Res 44(4):1027–1042

Pal S, Mukherjee SN, Ghosh S (2014b) Application of HYDRUS 1D model for assessment of phenol–soil adsorption dynamics. Environ Sci Pollut R 21:5249–5261

Study on Photocatalyst for Removal of Organic Pollutants from the Water of Damodar River, India

Kankana Seal, Hirok Chaudhuri, Soumen Basu, Mrinal Kanti Mandal, Supriya Pal, Kashyap Kumar Dubey, and Pankaj Kumar Roy

1 Introduction

Different techniques such as aeration, softening, coagulation, flocculation, adsorption, de-infection, microfiltration membrane technology, ultrafiltration and nanofiltration membrane technique, reverse osmosis membrane process, ion exchange, carbon adsorption techniques, bio-remediation and many more have been reported by several researchers to remove the pollutants from water bodies (El-ghonemy 2013; Pc'rcz-Gonza et al. 2012; Djilani et al. 2012). Moreover, phase separation, sedimentation, filtration, biochemical oxidation, and chemical oxidation are well-known techniques for treating wastewater to remove organic and inorganic contaminants. However, none of these techniques is more effective for the total mineralisation of

K. Seal
Department of Chemical Engineering, Indian Institute of Technology - Indian School of Mines Dhanbad, Dhanbad, India

H. Chaudhuri (✉) · S. Basu
Department of Physics, National Institute of Technology Durgapur, Durgapur, India
e-mail: hchaudhuri.phy@nitdgp.ac.in

H. Chaudhuri
Center for Research On Environment and Water (CREW), National Institute of Technology Durgapur, Durgapur, India

M. K. Mandal
Department of Chemical Engineering, National Institute of Technology Durgapur, Durgapur, India

S. Pal
Department of Civil Engineering, National Institute of Technology Durgapur, Durgapur, India

K. K. Dubey
School of Biotechnology, Jawaharlal Nehru University, New Delhi, India

P. K. Roy
School of Water Resources Engineering, Jadavpur University, Kolkata, India

Y. Li et al. (eds.), *BRICS Countries: Sustainable Water Resource Management and Pollution Control*, https://doi.org/10.1007/978-981-99-9581-3_5

toxic organic chemicals (Tisa et al. 2014) in contaminated water. Advanced Oxidation Processes (AOP) is appeared to be the most established method to degrade chemically stable organic pollutants in wastewater (Chou and Huang 1999. In the present work, we have focused our attention on the fabrication of heterogeneous photocatalyst using AOP to remove organic pollutants from wastewater. The present paper is focused on developing sustainable technology to remove organic pollutants from the Damodar river water flowing in the eastern part of India. Phenol content is very high in the water of the Damodar river (0.85 ppm) as it is highly polluted by the effluents of the steel plants situated on the bank of Damodar in the Asansol-Durgapur industrial zone. In the present work, methylene blue (MB) has been chosen as a model organic pollutant as it is one of the most stable cationic dyes. This research work aims to degrade phenol and its derivatives (di-tert-butyal ethyl phenol, nonyl phenol, bisphenol A, nitrophenol, etc.) from water bodies. Various photocatalysts with different nanomaterial structures, such as nanotubes, nanorods, microspheres, and even nano-flower, have been synthesised by numerous researchers to degrade organic contaminants from water bodies (Gaya 2014; Herrmann et al. 2002). In recent times, metal oxide nanomaterial has shown extensive applications as a photocatalyst in removing organic contaminants from water bodies due to its controlled phase, dimension, morphology, size, and physicochemical properties (Hoffmann et al. 2002). Some metal oxide nanomaterial like TiO_2 falls in the semiconductor group, showing photocatalyst activity (Gaya 2014). Long-term chemical stability in water media, cost-effectiveness, and reusability of Titanium-based oxide materials makes them important photocatalysts (Herrmann et al. 2002). This photocatalyst is governed by faster photon-induced electron transportation (Herrmann et al. 2002). In nature, TiO_2 is found in three crystalline phases such as rutile (tetragonal), anatase (tetragonal structure), and brookite (orthorhombic). Remarkably, anatase shows the highest photocatalytic behaviour compared to the other polymorphs of titania (Hanaor and Sorrell 2011). Moreover, the efficiency of the photocatalytic activity of TiO_2 is governed by several physical properties such as morphology, small crystalline size, porosity, the ratio of anatase to rutile phase, pore volume, specific surface area, quantum efficiency, etc. (Herrmann 2010a). The photocatalytic activity of TiO_2-based materials can be modified in terms of adaptation of phase, dimensions, different structural morphology, crystalline size, surface area, etc. (Herrmann 2010b; Linsebigler et al. 1995; Fujishima et al. 2000). Because of the application-oriented requirement, TiO_2 nanostructures can be synthesised in the forms of spheres, rods, sheets, wires, belts, etc. (Walaa et al. 2019; Fernandez et al. 1995; Lin et al. 1995) using various experimental methods such as sol–gel, electrochemical anodisation, hydrothermal/solvothermal, electrodeposition, and several template-assisted routes (Chen and Mao 2007; Wang et al. 2011; Ghicov et al. 2006; Etacheri et al. 2011; Shen et al. 2011; Lai et al. 2009; Mor et al. 2006; Ghicov and Schmuki 2006; Kasuga et al. 1998; Bavykin et al. 2006; Shi et al. 2013).

The present work explores the research activity performed towards synthesising mesoporous TiO_2 nanostructure-based photocatalyst by the solvothermal method using the AOP technique. The AOP methods include ozonation (Pilli et al. 2015), fenton reactions (Rahim Pouran et al. 2015; Magureanu et al. 2015), plasma oxidation (Xin et al. 2016; Shu et al. 2016), UV/H_2O_2 (Saien et al. 2012; Devi and Kavitha

2013), photocatalytic reactions (Kumar and Devi 2011; Oller et al. 2011). These methods deal with the generation of hydroxyl radicals (·OH), the strongest oxidising agent with a redox potential of 2.8 V (García-Montaño et al. 2008) that helps break the stabilised organic compound in aqueous media. In this context, it has been observed that TiO_2-based material has become the most popular heterogeneous photocatalyst via photocatalytic reactions due to its long-term stability, inexpensiveness, and non-toxicity (Cortez et al. 2010). In TiO_2-based material, the energy difference between the valence band and the conduction band, called band gap energy, is around 3.2 eV. Irradiation of light (photon) with higher energy than band gap energy can be absorbed by the photocatalyst leading to the excitation of the electrons from the valence band to its conduction band. After absorption of photon energy-rich electrons (e_{CB}^-) are gathered in the conduction band, and holes ($h\nu_B^+$) are generated in valence bands (Wang et al. 2011), and as a result, electron–hole pairs are generated in the valence band as shown in Eq. (1). $h\nu_B^+$ is a powerful oxidising agent and can directly degrade the stable structure of organic compounds resulting in the formation of CO_2 and other by-products shown in Eq. (2). As shown in Eq. (3) $h\nu_B^+$ also reacts with adsorbed water molecules on the catalyst surface, resulting in one of the most powerful reactive species ·OH (hydroxyl radical). ·OH can easily oxidise organic compounds in water media due to its second-highest oxidation potential presented in Eq. (4).

$$TiO_2 + h\nu \rightarrow e_{CB}^- + h_{VB}^+ \tag{1}$$

$$h_{VB}^+ + R \rightarrow CO_2 + H_2O + \text{By - products} \tag{2}$$

$$H_2O + h_{VB}^+ \rightarrow H_2O+ \rightarrow \cdot OH + H^+ \tag{3}$$

$$\cdot OH + R \rightarrow \text{intermediate products} \rightarrow CO_2 + H_2O \tag{4}$$

Nevertheless, superoxide radical $O_2^{\cdot -}$ is generated by the reaction between e_{CB}^- with O_2 shown in Eq. (5). This superoxide radical may also participate in the degradation process.

$$e_{CB}^- + O_2 \rightarrow O_2^{\cdot -} \tag{5}$$

Figure 1 demonstrates the formation of electron–hole pair and •OH radicals.

In the present work, we have used $TiCl_3$ precursor. Herein, we have varied the solvothermal reaction temperature to observe its effect on the morphology, phase, specific surface area, and other physicochemical properties of the prepared TiO_2-based photocatalyst and its photocatalytic efficiency. The present paper also includes the details of the experimental procedures through XRD, FESEM, BET surface analysis, UV–Vis spectroscopy, etc. Details of this different characterisation have been discussed in this paper. Furthermore, optimum characteristics, i.e. morphological structure, phase composition, existing functional groups, surface area, etc., of the

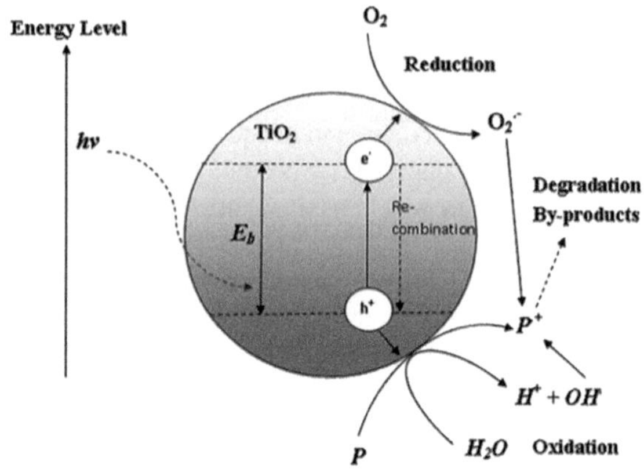

Fig. 1 Mechanism of photocatalytic degradation on TiO_2 material

synthesised titania-based photocatalysts have been suggested in this work regarding the removal of model organic pollutant methylene blue with its highest efficiency.

2 Experimental Details

2.1 Materials and Methods

$TiCl_3$ was taken as a titanium precursor to synthesise the hierarchical TiO_2 nano-flower. Absolute ethanol was taken as a solvent. The pH of the solution was controlled by sodium hydroxide. Subsequently, the mixture was transferred to an autoclave and left for 18 h in an oven for solvothermal treatment at 140 °C temperature, and the synthesised sample was named MESO-140. The obtained precipitate was rinsed with a mixture of distilled water and ethanol several times. Afterwards, the same was dried at 50 °C for 2 h and calcined at 400 °C for 2 h to obtain a hierarchically structured TiO_2 nano-flower in powder form. The exact route of synthesis was repeated at different reaction temperatures, such as 150 and 160 °C. The synthesised samples at those temperatures were defined as MESO-150 and MESO-160, respectively.

2.2 Characterisation

X-ray diffractometer was used to carry out X-ray diffraction (XRD) experiments with a variation in 2θ (10°–80°). Average crystalline size (D) was estimated by the well-known Debye-Sherrer equation (Danilchenko et al. 2002; Langford and Wilson 1978]. Field Emission Scanning Electron Microscopy (FESEM) was used to study the surface morphology of the synthesised material MESO-150. The surface area of the material was investigated through the Surface Area Analyser using N_2 adsorption–desorption method by means of Braunauer-Emmet-Teller (BET) equation (Fagerlund 1973; Lowell et al. 2004). Barret-Joynet-Halenda (BJH) formula (Klobes et al. 2006; Bardestan et al. 2019) using the desorption characteristics of the material was used for the determination of pore size distributions of the samples. The optical properties of the sample were investigated by means of UV–Vis absorbance spectroscopy. Notably, analysed samples were used in the form of fine powder.

2.3 Photocatalytic Activity Measurement

Photocatalytic degradation tests of methylene blue were performed to investigate the photocatalytic activity of the MESO-150 under both dark and UV irradiation conditions in an aqueous solution at ambient conditions (temperature 25 °C ± 4). The photocatalytic reactor was equipped with a closed black wooden box and four UV-C ($\lambda = 254$ nm; model Philips TUV-11W-G11-T5) lamps which were used as the source of UV irradiation. Here adsorption–desorption equilibrium of methylene blue on the surface of the samples in a water medium was studied. 40 ppm concentrated methylene blue solution (at pH = 7) was taken into a measuring beaker without any catalyst loading for the above-said purpose. The solution was kept in the dark conditions with continuous magnetic stirring for 30 min until the equilibrium of adsorption–desorption was reached. Afterwards, this adsorption–desorption phenomenon was confirmed by analysing a certain amount of methylene blue solution. Henceforth 0.5 gm/L amount of synthesised TiO_2 catalyst loading was made on the same solution of methylene blue. Similarly, the solution was continuously stirred through a magnetic stirrer for 120 min in dark conditions and adsorption of dye molecules was observed on the catalyst's surface. The experiment continued up to the surface adsorption–desorption equilibrium. The same procedure was performed for 200 min under UV illumination to study the photocatalytic activity of titania samples. Afterwards, a certain amount of methylene blue solution (20 ml) was collected in a fixed time interval and was filtered through the syringe filter to examine the kinetics of individual photocatalytic degradation reactions. Each collected sample was analysed using UV–Visible Spectrophotometer (at wavelength 665 nm) to observe the time-wise photocatalytic disappearance of methylene blue solution. Generally, for the photocatalytic reaction, the rate constant is used to describe how fast the photocatalytic degradation reactions occur using each of the prepared samples. Practically, heterogeneous photocatalysis (using synthesised TiO_2) follows the Langmuir-Hinshelwood mechanism, and the rate of the reaction (rt) is proportional to the coverage (θ) of reactants in the following way,

$r_t = k_t\theta = \frac{k_t K_a C}{1+K_a C}$ (Herrmann 2010a), where k_t is the true rate constant in specific reaction conditions such as temperature, the concentration of the pollutant (C), the mass of the catalyst, pH, light intensity, etc., and K_a is the adsorption constant. Langmuir- Hinshelwood mechanism remains valid when and only when the initial concentration of the reactants is very low (Herrmann 2010a, b) (here, initial concentration $C_0 = 40$ mg/l) and the denominator term K_aC in Langmuir- Hinshelwood mechanism can be neglected. In that scenario, the rate constant may be expressed as $r_t = \frac{-dC}{dt} = k_tK_aC = kC$ (Herrmann 2010a, b), where k denotes the apparent pseudo-first-order rate constant. Therefore, the rate equation now becomes $\ln \frac{C_0}{C_t} = kt$, where C_t = concentration of methylene blue at a certain irradiation time (t). Based on the equation, the rate constants were estimated through the slope of the irradiation time (t) versus $\ln (C_0/C_t)$ plot. The percentage degradation of methylene blue was calculated by means of $\frac{C_t - C_0}{C_0} \times 100\%$. Similarly, surface adsorption and photocatalytic activity tests were repeated using the different TiO_2 samples prepared at various solvothermal temperatures.

3 Results and Discussions

3.1 *Crystal Structure Analysis (XRD Analysis)*

Figure 2 shows X-ray diffraction patterns of the synthesised TiO_2 materials. In the present work, observed diffraction peaks for anatase phase (at $2\theta = 25.32^o$) and rutile phase (at $2\theta = 27.37°$) for the crystallographic planes (101) (100% intensity) and crystallographic plane (110) (100% intensity) respectively (Fig. 2). Figure 2 shows that the low reaction temperature, i.e. at 140 ºC rutile phase (denoted as R) becomes more prominent than anatase (denoted as A) (i.e. in MESO-140). With a further increment of reaction temperature (i.e. MESO-150 and MESO-160), anatase peak intensities become stronger, and the width of the observed peaks appears to be narrower, as shown in Fig. 1 due to the fast rate of nucleation reaction at a high reaction temperature. It generally depicts about enhancement of crystallisation in all synthesised TiO_2 samples according to the rise of solvothermal temperature. The average crystalline sizes of all prepared samples have been calculated using Scherrer's equation, as tabulated in Table 1.

Moreover, variations in relative intensities of anatase and rutile contents were observed among all prepared MESO samples as shown in weight fraction in **Table 1** (using the concept as adopted by Zhang and Banfield (2000); Zhang and Banfield 2000).

3.2 *Morphological Studies (SEM Analysis)*

The morphological structures of TiO_2 samples are highly impacted by solvothermal temperature. FESEM image of TiO_2 samples prepared at 150 °C reaction temperature is shown in Fig. 3. Hierarchical nano-flower along with nanorod spikes texture and

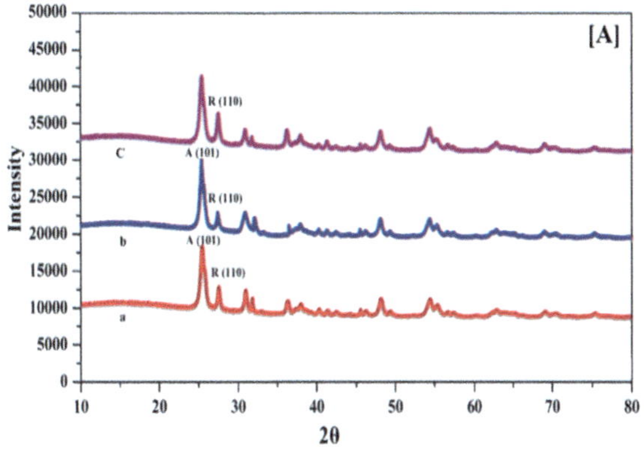

Fig. 2 XRD pattern of **a** MESO-140, **b** MESO-150 and **c** MESO-160

Table 1 The weight fraction of anatase phase content and crystalline size of MESO- 140, MESO-150, and MESO-160

Sample	Weight fraction of anatase	Crystalline size (nm)
MESO-140	0.72	11.69
MESO-150	0.81 (highest)	5.54 (lowest)
MESO-160	0.67	8.76

mesopores was observed. Numerous hierarchical units of spherical nano-flowers seem to be distributed uniformly for MESO-150 (Fig. 3).

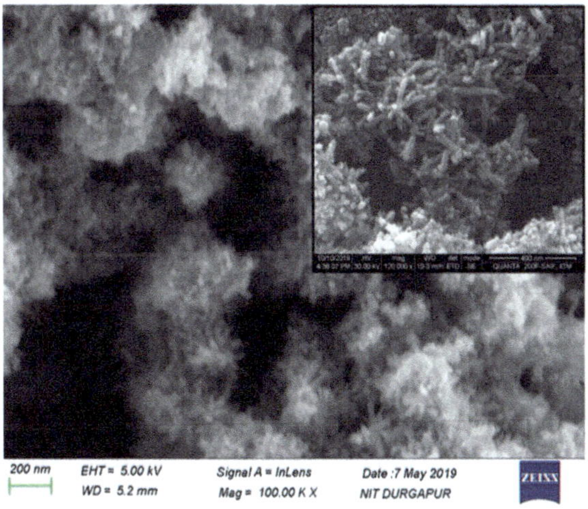

Fig. 3 FESEM images of MESO-150

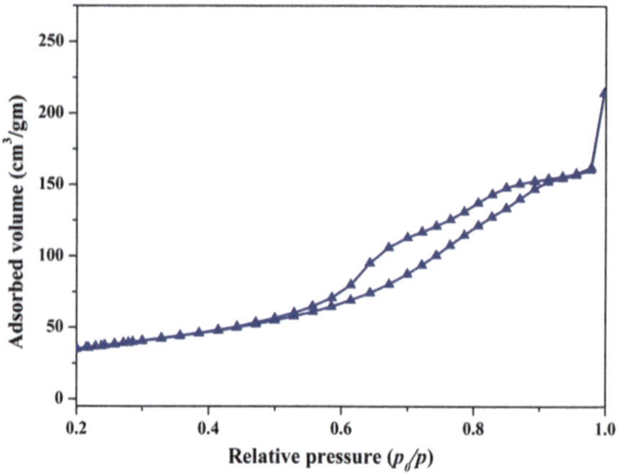

Fig. 4 N_2 adsorption–desorption BET isotherm of MESO-150 sample

Table 2 Surface area, pore volume, pore-size characteristics, particle size and band gap energy of TiO_2 sample (MESO-150)

Sample	Surface area (m^2g^{-1})	Pore size (nm)	Pore volume ($cm^3\ g^{-1}$)	Particle size (nm)	Surface to volume ratio	Band gap (eV)
MESO-150	126.14	10.04	0.31	47.56	406.90	2.93

3.3 *Surface Area Analysis—N_2 Adsorption–Desorption Isotherm (BET Analysis)*

Nitrogen adsorption–desorption isotherm was used to investigate the surface area characteristics and thermal stabilities of MESO-150 as shown in Fig. 4. Hysteresis loop (type IV) seems to be formed in the region around $0.45 < p/p0 < 0.85$ and that falls within type H_3 and H_2. According to the type of hysteresis loops of MESO-150, the pore characteristics seem like large cavities with narrow entrances, and it appears to be an ink-bottle-type structure combined with slit-like particles. The average particle size of prepared samples turns out to be 47.56 nm, and the surface area is found to be 126.14 $m^2\ g^{-1}$ (Table 2). Observation shows that a reaction temperature of 150 °C appears to be the intermediate reaction temperature to achieve the sample's largest specific surface area and smallest particle size.

3.4 *Pore-Size Distribution (BJH Analysis) Analysis and Surface-To-Volume Ratio*

BJH techniques and the desorption characteristics were used to estimate pore-size distribution. Table II reflects that the pore size falls under the characteristics of mesopores adopted by the BDDT classification, as explained by Wang et al. (2012). Here, dual pore-size characteristics, i.e. bimodal pore distribution, were observed in the samples. The bimodal pore distribution consists of a combination of two types of pores, i.e. (i) tiny intra-aggregated pores and (ii) large inter-aggregated pores, as discussed by Yu et al. (2007) and this pore-characteristics was found in sample MESO-150. The pore volume of MESO-150 seems to be 0.31 cm^3 g^{-1} (Table 2). In addition, the probability of achieving more reactive sites to involve reactants in aqueous media can be enhanced with a high specific surface area to pore volume ratio for a sample. Therefore, the high value of specific surface area per unit volume is strongly needed to achieve a high catalytic reaction rate. MESO-150 exhibits the highest surface area to volume ratio of all the samples, around 406.90.

3.5 UV–Vis Absorbance Spectroscopy Study

Figure 5 shows UV–Vis absorbance spectra which reflect the optical property of the MESO-150 sample. The absorption band was observed in the UV region ~350 nm. There was moderate UV absorption found in the case of MESO-150 compared to the rest of the samples. Kubelka–Munk function was used to calculate the band gap energy by making a plot of photon energy (hν) versus $(F(R)\ h\nu)^{1/2}$ for the prepared MESO-150 sample (Escobedo-Morales et al. 2019). The estimated band gap energy for MESO-150 is 2.93 eV (Table 2). Based on all the facts mentioned above, it is demonstrated that the effect of reaction temperature plays a significant role in achieving the necessary UV absorption capacity and the band gap energy. This turn, directly affects the photocatalytic activity.

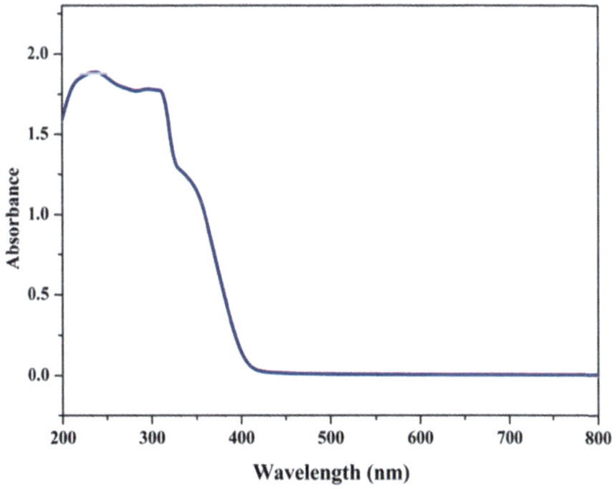

Fig. 5 UV–Vis Absorbance spectrum of prepared MESO-150 sample

3.6 Photocatalytic Activity

The performance of photocatalytic activity is controlled by existing phase content, morphological structure, particle size, pore size, specific surface area, etc. (Okeke et al. 2013; Cheng et al. 2014). Photocatalytic activities of MESO-150, MESO-160 and MESO-180 were studied through the degradation of methylene blue (MB) solution by monitoring its absorbance under dark conditions and UV radiation. Relative changes in concentration, i.e. C_t/C_0 of MB were assessed at different time intervals, as shown in Fig. 6. Adsorption of methylene blue in water media without contact to any light source is almost negligible. In contrast, a very small fraction of adsorption of MB dye molecules over the catalyst surface was observed after the incorporation of optimal titania catalysts. The sample MESO-150 shows the highest adsorption capacity of MB on the catalyst's surface and shows the best performance in MB degradation under UV irradiation due to its high specific surface area compared to other samples.

The performance (percentage of degradation) of photocatalytic degradation of MB solution using synthesised hierarchical TiO_2 nano flower is shown in Table 3. The value of regression coefficient R^2 was estimated for MESO-150 from the regression analysis of the kinetics study, as shown in Table 3. MESO-150 exhibits the fastest rate of reaction, around 0.0287 min^{-1}.

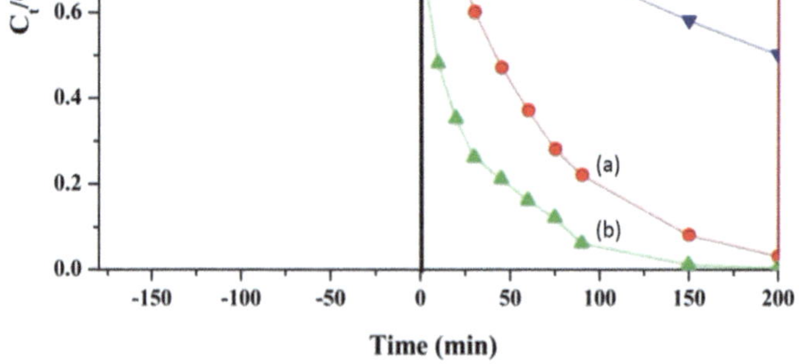

Fig. 6 Relative change in concentration of methylene blue dye in dark conditions and UV irradiation using titania samples (**a** MESO-140, **b** MESO-150, and **c** MESO-160)

Table 3 Parameters for Pseudo first-order kinetics of TiO_2 sample (MESO-150)

Sample	Percentage of degradation (%)	Rate of degradation (min^{-1})	R^2
MESO-150	99.84	0.0287	0.99

4 Conclusion

$TiCl_3$ as titanium precursors were used to synthesising hierarchical TiO_2 nano flower structures. The influence of solvothermal temperature on hierarchical TiO_2 nano flower was studied to optimise the most favourable temperature in which the synthesised material will have high photocatalytic efficiency. It was found that the reaction temperature strongly affects the phase content, morphological structures, crystallinity, particle size, specific surface area, pore-size distribution, surface-to-volume ratio,

optical properties, etc. Results indicate that the TiO_2 nanostructure, which was synthesised at a reaction temperature of 150 °C, only shows high catalytic efficiency. XRD analysis and FESEM of the sample MESO-150 exposed numerous organised hierarchical structures with high crystalline anatase phase content. High specific surface area and moderate average pore size were obtained at a solvothermal temperature of 150 °C which was confirmed by BET and BJH analysis. The signature of bimodal pore-size distribution in MESO-150 samples provides the property of high-rate reaction kinetics in MESO-150, and those were missing in other samples. The adopted route of synthesis is considered to be one of the simplest methods to develop mesoporous titania-based photocatalysts via the hydrolysis reaction of $TiCl_3$. Based on the present study, the degradation of phenol and its derivatives will be observed using the synthesised MESO-150 nanostructure in the near future to remove the phenol content of the river Damodar. Synthesised hierarchical TiO_2 photocatalysts could be useful in the field of wastewater treatment.

Acknowledgements The authors would like to acknowledge NIT Durgapur for financial and other administrative support to conduct the research. The authors extend their heartiest thanks to the funding agency DST Govt. of India for supporting through project under BRICS Multilateral Call 2017 (Project Grant No. DST/IMRCD/Pilot Call 2/Enviorganic/2018 (G) dated 28.03.2019).

References

Bardestan R, Patience SG, Kaliaguine S (2019) Experimental methods in chemical engineering: specific surfacearea and pore size distribution measurements—BET, BJH, and DFT. Can J Chem Eng 97:2781–2791

Bavykin DV, Friedrich JM, Walsh FC (2006) Protonated titanates and TiO_2 nanostructured materials: synthesis, properties, and applications. Adv Mater 18:2807–2824

Chen X, Mao SS (2007) Titanium dioxide nanomaterials: synthesis, properties, modifications, and applications. Chem Rev 107:2891–2959

Cheng H, Wang J, Zhao Y, Han X (2014) Effect of phase composition, morphology, and specific surface area on the photocatalytic activity of TiO_2 nanomaterials. RSC Adv 4:47031–47038

Chou S, Huang C (1999) Application of a supported iron oxyhydroxide catalyst in oxidation of benzoic acid by hydrogen peroxide. Chemosphere 38(12):2719–2731

Cortez S, Teixeira P, Oliveira R, Mota M (2010) Evaluation of Fenton and ozone-based advanced oxidation processes as mature landfill leachate pre-treatments. J Environ Manage 92(3):749–755

Danilchenko SN, Kukharenko OG, Moseke C, Protsenko IY, Sukhodub LF, Sulkio-Cleff B (2002) Determination of the bone mineral crystallite size and lattice strain from diffraction line broadening. Cryst Res Technol J Exp Indus Crystallogr 37:1234–1240

Devi LG, Kavitha R (2013) A review on non metal ion doped titania for the photocatalytic degradation of organic pollutants under UV/solar light: role of photogenerated charge carrier dynamics in enhancing the activity. Appl Catal B: Environ 140–141:559–587

Djilani C, Zaghdoudi R, Modarressi A, Rogalski M, Djazi F, Lallam A (2012) Elimination of organic micropollutants by adsorption on activated carbon prepared from agricultural waste. Chem Eng J 189–190:203–212

El-ghonemy AMK (2013) Waste energy recovery in seawater reverse osmosis desalination plants. Part 1. Rev Renew Sustain Energy Rev 18:6–22

Escobedo-Morales A, Ruiz-López II, de Ruiz-Peralta ML, Tepech-Carrillo L, Sánchez-Cantú M, Moreno-Orea JE (2019) Automated method for the determination of the band gap energy of pure and mixed powder samples using diffuse reflectance spectroscopy. Heliyon 5:1–19

Etacheri V, Seery MK, Hinder SJ, Pillai SC (2011) Oxygen rich titania: a dopant free, high temperature stable, and visible-light active anatase photocatalyst. Adv Func Mater 21:3744–3752

Fagerlund G (1973) Determination of specific surface by the BET method. Matériaux Et Construct 6:239–245

Fernandez A, Lassaletta G, Jimenez VM, Justo A, Gonzalez-Elipe AR, Herrmann JM, Tahiri H, Ait-Ichou Y (1995) Preparation and characterisation of TiO_2 photocatalysts supported on various rigid supports (glass, quartz and stainless steel). Comparative studies of photocatalytic activity in water purification. Appl Catal B 7:49–63

Fujishima A, Rao TN, Tryk DA (2000) Titanium dioxide photocatalysis. J Photochem Photobiol C 1:1–21

García-Montaño J et al (2008) Pilot plant scale reactive dyes degradation by solar photo-Fenton and biological processes. J Photochem Photobiol A: Chem 195(2):205–214

Gaya IU (2014) Heterogeneous photocatalysis using inorganic semiconductor solids. Springer Dordrecht Heidelberg London New York, pp 1–16. ISBN: 978-94-007-7775-0

Ghicov A, Macak JM, Tsuchiya H, Kunze J, Haeublein V, Frey L, Schmuki P (2006) Ion implantation and annealing for an efficient N-doping of TiO_2 nanotubes. Nano Lett 6:1080–1082

Ghicov A, Schmuki P (2006) Self-ordering electrochemistry: a review on growth and functionality of TiO_2 nanotubes and other self-aligned MOx structures. Chem Commun 2791–2808. https://doi.org/10.1039/B822726H

Hanaor DA, Sorrell CC (2011) Review of the anatase to rutile phase transformation. J Mater Sci 46:855–874

Herrmann JM (2010a) Photocatalysis fundamentals revisited to avoid several misconceptions. Appl Catal B 99:461–468

Herrmann JM (2010b) Fundamentals and misconceptions in photocatalysis. J Photochem Photobiol, A 216:85–93

Herrmann JM, Guillard C, Disdier J, Lehaut C, Malato S, Blanco J (2002) New industrial titania photocatalysts for the solar detoxification of water containing various pollutants. Appl Catal B 35:281–294

Hoffmann MR, Martin ST, Choi W, Bahnemann DW (2002) Environmental applications of semiconductor photocatalysis. Chem Rev 95:69–96

Kasuga T, Hiramatsu M, Hoson A, Sekino T, Niihara K (1998) Formation of titanium oxide nanotube. Langmuir 14:3160–3163

Klobes P, Meyer K, Munro GR (2006) Porosity and specific surface area measurements for solid materials. National Institute of Standard and Technology, Spec. publ. 960–17, Washington

Kumar SG, Devi LG (2011) Review on modified TiO_2 photocatalysis under UV/visible light: selected results and related mechanisms on interfacial charge carrier transfer dynamics. J Phys Chem A 115(46):13211–13241

Lai Y, Gao X, Zhuang H, Huang J, Lin C, Jiang L (2009) Designing superhydrophobic porous nanostructures with tunable water adhesion. Adv Mater 21:3799–3803

Langford JI, Wilson AJC (1978) Scherrer after sixty years: a survey and some new results in the determination of crystallite size. J Appl Crystallogr 11:102–113

Lin J, Heo YU, Nattestad A, Sun Z, Wang L, Kim JH, Dou SX (1995) 3D hierarchical rutile TiO_2 and metal-free organic sensitiser producing dye-sensitised solar cells 8.6% conversion efficiency. Sci Rep 4:1–8

Linsebigler AL, Lu G, Yates JT Jr (1995) Photocatalysis on TiO_2 surfaces: principles, mechanisms, and selected results. Chem Rev 95:735–758

Lowell S, Shields JE, Thomas MA, Thommes M (2004) Characterisation of porous solids and powders: surface area, pore size and density, 4th edn. Springer, Dordrecht, pp 58–81

Magureanu M, Mandache NB, Parvulescu VI (2015) Degradation of pharmaceutical compounds in water by non-thermal plasma treatment. Water Res 81:124–136

Mor GK, Shankar K, Paulose M, Varghese OK, Grimes CA (2006) Use of highly-ordered TiO_2 nanotube arrays in dye-sensitised solar cells. Nano Lett 6:215–218

Okeke G, Hammond RB, Antony SJ (2013) Influence of size and temperature on the phase stability and thermosphysical properties of anatase TiO_2 nanoparticles: molecular dynamics simulation. J Nanopart Res 15:1–9

Oller I, Malato S, Sánchez-Pérez JA (2011). Combination of advanced oxidation processes and biological treatments for wastewater decontamination—a review. Sci Total Environ 409:4141–4166

Pérez-González A, Urtiaga AM, Ibáñez R, Ortiz I (2012) State of the art and review on the treatment technologies of water reverse osmosis concentrates. Water Res 46:267–283

Pilli S et al (2015) Overview of Fenton pre-treatment of sludge aiming to enhance anaerobic digestion. Rev Environ Sci Biotechnol 14(3):453–472

Rahim Pouran S, Abdul Aziz AR, Wan Daud WMA (2015) Review on the main advances in photo-Fenton oxidation system for recalcitrant wastewaters. J Indus Eng Chem 21:53–69

Salen J, Moradi V, Soleymani A-R (2012) Investigation of a jet mixing photo-reactor device for rapid dye discoloration and aromatic degradation via UV/H_2O_2 process. Chem Eng J 183:135–140

Shen J, Zhu Y, Chen C, Yang X, Li C (2011) Facile preparation and upconversion luminescence of graphene quantum dots. Chem Commun 47:2580–2582

Shi H, Chen J, Li G, Nie X, Zhao H, Wong PK, An T (2013) Synthesis and characterisation of novel plasmonic Ag/AgX-CNTs (X= Cl, Br, I) nanocomposite photocatalysts and synergetic degradation of organic pollutant under visible light. ACS Appl Mater Interfaces 5:6959–6967

Shu H-Y, Chang M-C, Hsieh W-P (2016) Remedy of dye manufacturing process effluent by UV/H_2O_2 proces. J H Mat 28(1):60–66

Tisa F, Abdul Raman AA, Wan Daud WMA (2014) Applicability of fluidised bed reactor in recalcitrant compound degradation through advanced oxidation processes: a review. J Environ Manage 146:260–275

Walaa AA, Ibrahim HA, Basant AA, Nashaat A, Mohamed MA, Marwan YR, Noha I, Mona AM, Nageh KA (2019) Recent advances in the use of TiO_2 nanotube powder in biological, environmental, and energy applications. Nanoscale Adv 1:2801–2816

Wang D, Liu L, Zhang F, Tao K, Pippel E, Domen K (2011) Spontaneous phase and morphology transformations of anodised titania nanotubes induced by water at room temperature. Nano Lett 11:3649–3655

Wang W, Liu P, Zhang M, Hu J, Xing F (2012) The pore structure of phosphoaluminate cement. Open J Compos Mater 02:104–112

Xin L et al (2016) Degradation of triclosan in aqueous solution by dielectric barrier discharge plasma combined with activated carbon fibers. Chemosphere 144:855–863

Yu J, Wang G, Cheng B, Zhou M (2007) Effects of hydrothermal temperature and time on the photocatalytic activity and microstructures of bimodal mesoporous TiO_2 powders. Appl Catal B: Environ 69:171–180

Zhang H, Banfield JF (2000) Understanding polymorphic phase transformation behavior during growth of nanocrystalline aggregates: insights from TiO_2. J Phys Chem B 104:3481–3487

Integrated River Restoration Protocol–the Restoration of Lower Order Streams in Large River Networks

Vinod Tare and Gautam Roy

1 Introduction

Rivers are treasure troves of valuable resources (river ecosystem goods) for humans such as freshwater, food, and energy. They also serve numerous other human needs (river ecosystem services) such as the purification and drainage of anthropogenic wastewaters and flood waters and provide ready waterways for economic inland transport. Since the onset of industrialization and mega engineering projects, however, rivers have been exploited to the hilt in India, so that many of them have either ceased to flow altogether or to flow as moribund wastewater drains bereft of the ability to provide the multiple ecosystem services expected of rivers. Thus while rivers are highly dynamic ecosystems that operate on landscape scales, modern human actions can overwhelm their functioning over large spatiotemporal domains. River degradation in India has, therefore, had far-reaching and grave human impacts, with extreme floods, droughts, and diseases being prominent such disasters. They have also caused gradual erosion of the quality of human life in the river basin—from the loss of material and ecological benefits to those of cultural, aesthetic, and spiritual values in societies. River restoration and conservation is, hence, urgently needed in India to sustain its citizens and ecology for its present and future generations.

River restoration has been seriously taken up in India in recent decades, with a major focus on large iconic rivers such as the Ganga (or Ganges). A clear example of this is the preparation of a comprehensive Ganga River Basin Management Plan (GRBMP) in 2015 by a Consortium of 7 IITs (Indian Institutes of Technology) at the behest of the Indian government (Tare, Bose and Roy, 2015). But a River Management Plan, no matter how comprehensive it is, needs the resources and will to implement it. From a practical point of view, a fact that is often overlooked in

V. Tare (✉) · G. Roy
Centre for Ganga River Basin Management and Studies, Indian Institute of Technology Kanpur, Kanpur 208016, India
e-mail: vinod@iitk.ac.in

Y. Li et al. (eds.), *BRICS Countries: Sustainable Water Resource Management and Pollution Control*, https://doi.org/10.1007/978-981-99-9581-3_6

managing Indian rivers is that the source streams and lower order tributaries that bring in water, sediments, and nutrients—often from relatively un-impacted upland sources—are vitally important to maintain the health of large rivers (Wohl, 2017). They also contribute significant biodiversity to the river system. Thus, when river degradation begins in small tributaries, they call for immediate and focussed action to redress the problem and revive the low-order streams, without which the larger river's revival would remain incomplete and unstable.

2 The Focussed and Methodical Enterprise

In general, river restoration efforts in India have typically tried to revitalize large rivers without simultaneously reviving their smaller tributaries, but they have failed to achieve much success. This is due to the type, extent, and causes of degradations of large rivers being too diverse and distributed unevenly across the whole basin to be grasped in their entirety for river restoration to progress smoothly within limited time spans. Not only are such rivers too large to be revived and monitored thoroughly, but the human activities that adversely impact the river are themselves variable over time and space across the basin. A simpler approach to arrest and remedy the degradation process is to focus on smaller tributaries, especially those in urban areas that are perennial or those that can become perennial with treated wastewater inputs. Such tributaries include natural stormwater drains that now function mainly as wastewater drains or "nala"s in modern Indian cities. Many economic, environmental, aesthetic, and cultural benefits can accrue immediately for urban and peri-urban population groups by rejuvenating these small tributaries, and these early and tangible benefits can boost river rejuvenation efforts in the whole river basin. This is essentially a bottom-up approach that stabilizes the river restoration process as opposed to the top-down process. The latter frequently causes many stretches to remain vulnerable so that the river system as a whole remains prone to degeneration at critical times and locations. The process, as described by Tare et al. (2019), is illustrated in Fig. 1.

The meager and faltering achievements of river restoration efforts in India, however, are not only due to the enormity of the problem. Even when taken up in earnest on basin scales, restoration programs may be easily derailed by many vested interests involved in the programs. At the very outset, therefore, the restoration effort must clearly distinguish between the real stakeholders of the river from non-stakeholder players, and only the former should be actively engaged in the entire restoration program.

The lacuna in involving real stakeholders in river restoration in India may be because the river management authorities were themselves assigned an altogether different purpose than river restoration. Most river basin authorities in the country are effectively tasked with exploiting river resources and not about river conservation. Whereas their real task should focus on preserving the river ecosystem in the long term and ensuring that a river is not debilitated altogether or converted into a moribund drain. Nature and ecology have their own overarching importance which we may not

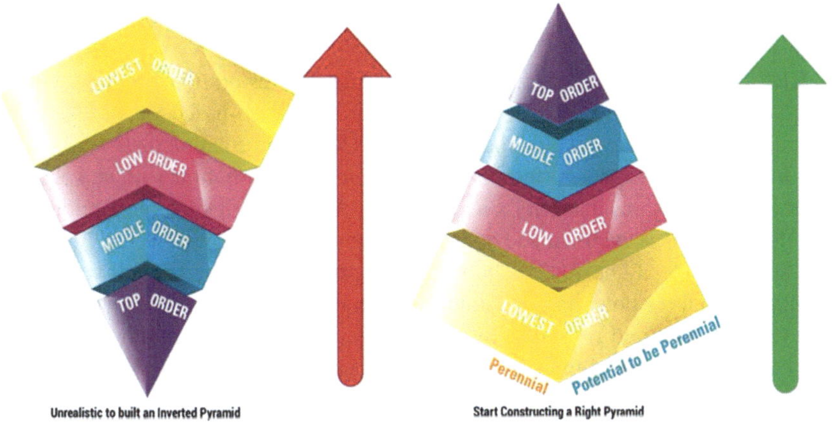

Fig. 1 The bottom-up approach to steadfast river restoration

comprehend from our short-term perspectives. River Basin Organizations (RBOs) must therefore embrace the supreme priority of river conservation above all other goals.

River conservation is often considered as anathema by development proponents. But once an RBO has adopted its primary goal as sustaining a river as river and the real stakeholders are fully involved in meeting this goal, social, and economic development in the river basin can accelerate rather than be hampered as happens at present. This is because the true sustainable benefits from the river–the Gross Ecosystem Product (Ouyang et al., 2020) of the river–can be availed systematically for steady development in the region instead of unsustainable extraction of one or two select resources leading to faltering growth. Availing a river's realistic benefits and providing the needed protective measures for the river are thus the main concerns of an RBO. River restoration and conservation can then be achieved without hiccups along with local and regional development.

RBOs must first gain stakeholders' confidence. Toward this end, RBOs must be committed to their prime goal of river restoration and conservation as well as being accountable to stakeholders. This will ensure full and fair implementation of the conservation measures in a timely manner. To give an example of why this is necessary, the present failure to protect rivers from pollution in India is seldom because pollution standards and waste disposal norms are deficient, but because of non-compliance with the standards due to deficient or even arbitrary policing by the government. But a set of vigilant stakeholders can resolutely track and prevent such harmful activities in the river basin. What they need in order to do this effectively is, first and foremost, scientific knowledge of rivers apart from being empowered to use that knowledge. And competent RBOs can impart scientific knowledge and understanding of river processes and river conservation to stakeholders. Hence, RBOs must themselves develop requisite knowledge and understanding of rivers under their charge, and discuss with and communicate to stakeholders what restoration measures are needed and why.

RBOs may also need to negotiate some of the desired restoration measures with the stakeholders themselves, and here again the hard evidence-based knowledge of the rivers can be crucial to get everyone on board. Hence information gathering, understanding of river processes, and communication and negotiation with stakeholders are the foremost steps that RBOs need to take. And through such interactions with stakeholders, RBOs would also begin to be accountable to them.

The above steps constitute a bottom-up approach whereby the rejuvenation of small perennial (or potentially perennial) urban rivers can pave the way for sustainable development and river restoration seamlessly all over the country. The entire task can be carried out in India by a network of independent RBOs for each degraded small tributary through information gathering, knowledge-building, widespread dissemination, and communication and negotiation with stakeholders as their basic tools.

3 What Should River Basin Organizations Strive for?

Keeping in view the local and global contexts of the tributary streams, the prioritized objectives of an RBO may be stated as below:

i. River Health Improvement to ensure that the river is restored as an ecologically stable and perennial river. This implies, that the following steps must be taken:

- Assured clean (i.e. uncontaminated) water inputs to the river in order to maintain the minimum desirable flows.
- Structural interventions like weirs and bunds to maintain adequate flow depth and width needed by aquatic flora and fauna.
- Debarring anthropogenic waste flows into the streams.
- Securing the physical integrity of rivers and their floodplains.

ii. Integration of the River with the Local Environment, especially:

- Hydraulic connectivity of the river with local water bodies.
- Unhampered natural drainage pathways in the catchment.
- Ensuring adequate natural vegetation and reducing built-up areas in the catchment.
- Moderating catchment runoff rates by vegetal measures to help increase soil moisture and recharge groundwater.

iii Maximizing Human Gains from River, for instance:

- Diversion and storage of surplus river water in nearby waterbodies for human needs.
- Cultivation of fish and other aquatic products in rivers and surface waterbodies.

- Promoting tourism and recreation.
- Development of navigation and waterway facilities in the region.
- Increasing government/municipal revenues from other river-based activities.

iv Curtailing Damages/Losses due to River, e.g.:

- Reducing the spread and durations of floods and water-logging through flood-plain zoning, drainage improvements, and vegetative protection measures.
- Reclaiming or converting water-logged areas for environmental improvements and urban uses.

v. Enhancing Benefits from River in Downstream Regions, such as:

- Better hydrological performance of downstream river stretches
- Improving water quality in downstream river stretches.
- Improved biodiversity in downstream river stretches.

4 Which Rivers Should Be Prioritized for Restoration

The tributaries that should be first selected for restoration should be medium-sized and relatively low-order streams that are not much affected by incoming tributaries that themselves need to be restored. The selected streams should not be too small and hydrologically insignificant; historically, they should also be perennial rivers–or at least those can be made to be perennial by additional locally available water inputs, since perennial streams can meet the restoration objectives reliably and ensure a definitive action plan for restoration. If, moreover, the selected streams are located in or near urban areas, then adequately treated municipal and industrial wastewaters, and irrigation return flows can easily augment the stream discharges. Selection of small to medium urban streams and natural drains has the advantage of not only restoring the streams easily but also of conveying treated wastewaters hygienically and aesthetically through the cities' waterways. The desired stream selection criteria can thus be summarized pointwise as below:

- The stream should be a small or medium-sized and relatively low-order river
- Historically it should have been a perennial stream, or one that can be made to flow round the year by locally available municipal water surpluses or treated, municipal, irrigation, or industrial wastewaters.
- The selected stream ought to be located in or around urban centers.
- It should not be a very small and insignificant river, but one that is relatively prominent and well-known in the region.

5 Governance of Rivers Under Restoration

To meet the river restoration objectives, the RBOs should adopt the following fundamental principles of governance:

- An RBO must, in principle, be custodian for the river.
- The RBO must adhere to stakeholder-based participatory governance.
- The RBO's actions must be in conformity with both State and Federal laws as well as any international agreements if applicable for the river.
- The RBO must engage in specialized and continuous knowledge updating.
- The RBO should maintain an open-access repository pertaining to all aspects and information about the river.

To fulfill the above principles, the RBO should comprise largely of eminent local people, community leaders, and stakeholders, besides government nominees and river experts.

6 Involving Stakeholders as Equal and Responsible Partners

Stakeholders of rivers are dependent on rivers in various ways, but they may not necessarily have emotional bonds with the rivers. In such cases or when they have other independent interests, their interests may or may not be aligned with river restoration goals. Hence, to bring all stakeholders to a common platform, stakeholders' understanding and knowledge about rivers and integrated water resource management (IWRM) may need to be taken up by RBOs. Figure 2 depicts the multiple aspects of such understanding and knowledge.

The above prerequisites can be best instilled in stakeholders through meaningful *communication and negotiation* by the RBOs. These actions broadly involve the following basic points:

(a) <u>Knowledge About and Understanding of River Processes</u>

  - Rivers as Dynamic Ecosystems (i.e., rivers flow continuously but variably; hence their characteristics may transform radically over time and distance along the river.)
  - Rivers as Constituent Elements Integrated within Water Resource Networks (i.e. rivers are hydraulically linked with floodplain wetlands, aquifers, and other waterbodies in the basin round the year or at least seasonally.)
  - Biogeochemical Water Quality Modulation and Nutrient Cycling in Rivers (i.e., the multiple bio-physical–chemical processes involving aquatic life, river flows, fluvial morphology, and basin geology that govern nutrient and material flows govern river water quality and the environment for aquatic life).

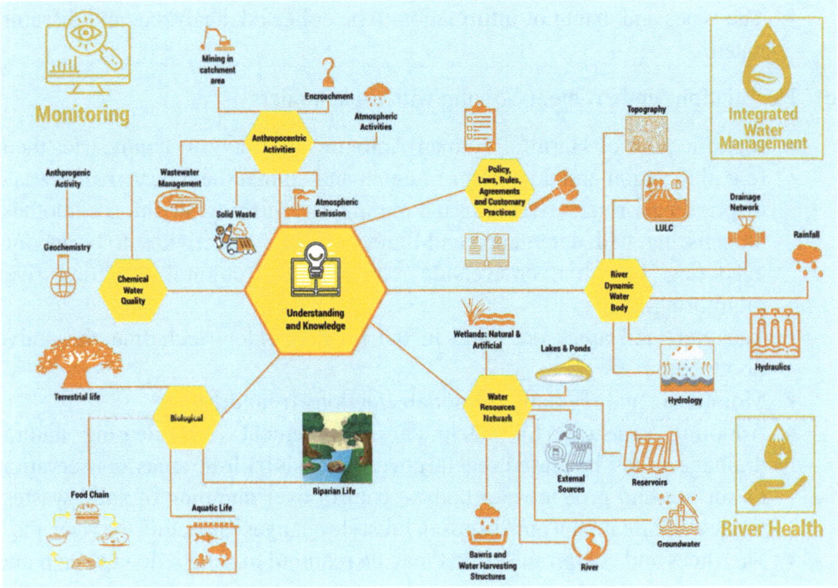

Fig. 2 Understanding and knowledge of river management and IWRM

- Benefits and Threats from Rivers (i.e. goods and services from the river ecosystem such as freshwater availability for domestic, agricultural, and industrial uses, fish and other biotic products for human needs, navigation, hydropower generation, tourism, and cultural benefits, and damages or disservices from river ecosystems like floods and barriers to mobility).
- In-stream and Off-stream Human Activities that affect rivers significantly
- River Health Indicators (to assess the status of Aquatic Biodiversity from visual observations and scientific measurements of river and riparian biota, hydraulic parameters, water quality, geomorphology, and other aspects).
- Formal/Legal Guidelines and Directives pertaining to Rivers (Policies, Laws and Rules of various National, State, and Local Authorities as well as international obligations, if any).

(b) Substantive Communication with Stakeholders

- All required information about rivers and knowledge of river processes.
- Local and regional importance of the river from social, economic, cultural, and ecological perspectives.
- Assessing/evaluating sustainable benefits from the river.
- Evaluating threats to the river from anthropogenic activities in the basin
- The fundamental role of stakeholders and their common or differentiated rights and responsibilities.
- Strategies and methods for planning, implementation, and monitoring of river restoration.

- The types and extent of information to be collected, analyzed, and disseminated.

(c) Negotiations and Problem-Solving with Stakeholders

- Identification of Harmful Human Activities and how to ban/restrict them (e.g. defecation and dumping of human and animal dung near rivers, waste disposal near rivers, riverbed sand mining, over-fishing, washing of clothes, utensils, etc. with detergents) and Beneficial Human Activities to be encouraged (e.g. benign river/riverside recreation and cultural activities, river monitoring, etc.)
- Vacating land areas occupied in the river corridor (including the active floodplains).
- Monitoring and regulating water abstractions from the river.
- Adopting measures for catchment improvement (e.g. clearing natural drainage routes, increased vegetal cover vis-a-vis built-up areas, conservation of surface and groundwater bodies, control over dumping of solid wastes, wastewater treatment and disposal, lifestyle changes impacting the river, etc.)
- Sacrifices and compromises that may be required to benefit downstream and upstream regions.

7 Restoration Strategy and Targets

The restoration strategy must cover comprehensive measures for the river as well as its floodplains and, in fact, the entire catchment. The various aspects that need to be considered include:

- Quality of Water Inflows into Rivers: Conventionally, municipal wastewaters are treated in Sewage Treatment Plants, whose treated discharges are rich in nutrients like Nitrogen and Phosphorous that can be very harmful to water bodies. A Four-Step Cycle to Improve the Water Quality to desired levels, as shown in Fig. 3, is therefore proposed for adoption for such wastewaters so that the receiving stream is ecologically and hygienically safe. As shown in the figure, the first two stages are Primary and Secondary Treatment of municipal sewage that remove most organic and inorganic components. In the third stage of treatment, wetlands (natural or constructed) phytoremediation removes most of the nutrients, residual organics, and other pollutants. The treated water reaching the river is thus virtually free of harmful elements while also meeting human needs and freshwater quality standards.

- Integrity of River: The river channel and riverbanks that form and stabilize naturally by hydrological-sedimentological processes over decades and centuries are highly vulnerable to indiscriminate human activities such as sand mining, and in-stream and riparian interventions. Such activities therefore need to be screened

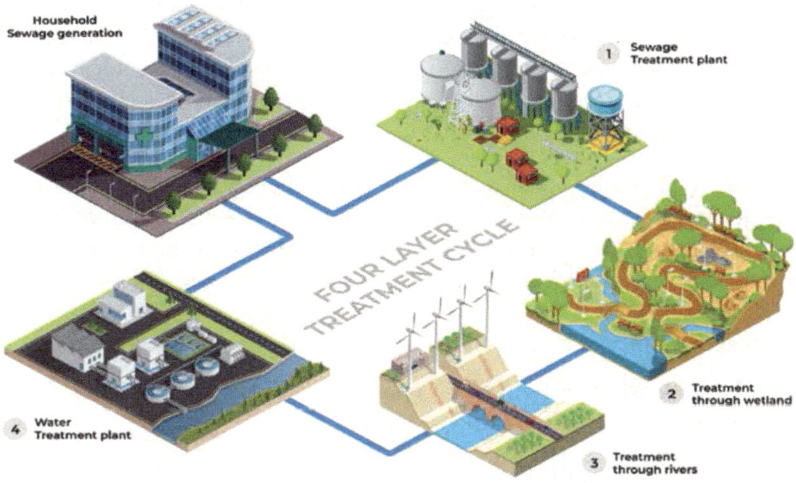

Fig. 3 The four-stage water quality improvement cycle

and strictly controlled. The structural integrity of, floodplains and the catchment as a whole including natural drains, waterbodies, vegetal cover, and drainage routes also need to be safeguarded against encroachments, impairments, obstructions, and blockages.

- Integrated Water Resource Network: Hydrologically rivers are freshwater bodies, but they seldom exist in isolation from other waterbodies; rather they are hydraulically and hydrologically connected to other neighboring water bodies. Preserving and developing the connectivity of catchment waterbodies (including groundwater) with the rivers is therefore both necessary and advantageous for regional water security.
- Infrastructure to make River Perennial and Eco-Friendly: Some structural changes may need to be introduced to ensure that the river flows perennially. For instance, weirs, especially in river stretches with steep gradients, would ensure desirable flow depths or flow velocities for river biodiversity. On the other hand, embankments may be desirable to prevent flooding, and bridges serve as river crossings for humans and land animals. But such interventions should be adopted keeping healthy river dynamics in focus.
- Conservation of Aquatic and Catchment Biodiversity: Flora and fauna of both rivers and their catchments are essential for healthy rivers. Natural vegetation in catchments helps in retaining soil moisture and soil biodiversity, enhancing groundwater recharge, moderating runoff rates, and purifying the runoff water. Hence natural vegetal cover in the basin needs to be maximized. Likewise, aquatic plants and animals should be maximized in the river and in waterbodies. Hence over-exploitation of aqua-cultural products must be checked and re-wilding with suitable species may be encouraged.

- Comprehensive Record Keeping: Alterations made and/or observed in the river and the river catchment from time to time and key monitoring indicators of the restoration-conservation period must be systematically recorded to assess the success of restoration efforts, help avoid repeated mistakes, and help guide restoration programs of other rivers.

Finally, it should be noted that all the goals of river restoration may not be achieved very soon in totality. The restoration targets should, therefore, be identified in sequential stages and milestones set out to achieve them. For instance, the expected achievement stages may be:

1st Stage–Aesthetic Satisfaction (based on visual appearance, odor, etc.).

2nd Stage–Satisfaction from Non-Contact Recreational Activities (e.g. boating and rafting).

3rd Stage–River Integrity (e.g. stable river bed and riverbanks, floodplains free of human obstructions) and Satisfactory Water Quality.

4th Stage–Reviving plentiful indigenous river biodiversity.

8 Monitoring and Evaluation, Feedback and Corrective Measures

Rivers being extremely dynamic and complex ecosystems, strategic planning with open-ended options are needed to remedy temporary setbacks in the restoration process. Hence stringent monitoring and evaluation requirements combined with feedback loops for corrective measures in the restoration plan are of the essence. The tasks involved may be briefly described as below:

- Engagement of independent specialized bodies to monitor both the Sewage Treatment Plants (STPs) as well as waterbodies (Wetlands, Drains) and the River being restored.
- Stakeholder feedback on efficacy of the restoration process and suggested measures for improvement.
- Performance evaluation and suggested measures for improvement by experts.
- Overall evaluation of the progress of restoration by the RBO.
- Review of all feedback and evaluation inputs and adoption of suitable corrective measures in Restoration Plan by the RBO.

9 Conclusion

India's rivers and other aquatic ecosystems have degraded significantly over many decades, and citizens concern about them has increased correspondingly. This concern has propelled many individuals, citizens' groups, and governments to address the problem in different ways but with limited effect so far. From these experiences over the last few decades and of river restoration and management efforts globally, a sound strategy has been presented here to restore degraded Indian rivers and freshwater bodies. This is essential for improving India's water security and environmental well-being for sustained urban, industrial, agricultural, and economic growth and public hygiene. It will also be a major step to integrate India's largely government-controlled water resource management activity with comprehensive, knowledge-based, and stakeholder-inclusive action.

References

Ouyang Z et al (2020) Using gross ecosystem product (GEP) to value nature in decision making. PNAS 117(25):14593–14601

Tare V, Roy G (2015) Ganga river basin management plan (GRBMP)–Volume 1: Main Plan Document. cGanga and NMCG.

Tare V et al. (2019) River restoration and conservation: A concise manual and guide. cGanga and NMCG.

Wohl E (2017) The significance of small streams. Front Earth Sci 11(3):447–456

China: Paths to Clean Water Under Increasingly Dynamic Environment

Yiping Li

1 The Present State of Water Pollution in China

1.1 Water Resources Overview in China

China's abundant freshwater resources total 2.8 trillion m^3, representing 6% of the global total and placing it 6th in the world, trailing Russia, Brazil, the USA, Canada, and Indonesia (Ministry of Water Resources of the People's Republic of China, 2015). Nevertheless, China's water resources per capita are relatively limited, standing at just 2,100 m^3, which is 28% of the global average (Fig. 1). This classification places China among the world's most water-scarce nations.

China's water resources face challenges concerning distribution and quality. Precipitation is unevenly distributed, with 60–80% occurring during the flood season, resulting in floods and droughts.

Additionally, the water resources spatial distribution does not align with productive forces and resource availability (Ministry of Water Resources of the People's Republic of China, 2015). Furthermore, China grapples with significant water pollution. According to the latest issue of Water Resources Bulletin of China, 78.5% of the total river reached Grade 1 and 3 water quality standards, while 8.3% fell below Grade 5 standards (Ministry of Water Resources of the People's Republic of China, 2015).

Y. Li (✉)
Department of Environment, University of Hohai, 1 XiKang RD, Gulou District, Nanjing, China
e-mail: liyiping@hhu.edu.cn

Y. Li et al. (eds.), *BRICS Countries: Sustainable Water Resource Management and Pollution Control*, https://doi.org/10.1007/978-981-99-9581-3_7

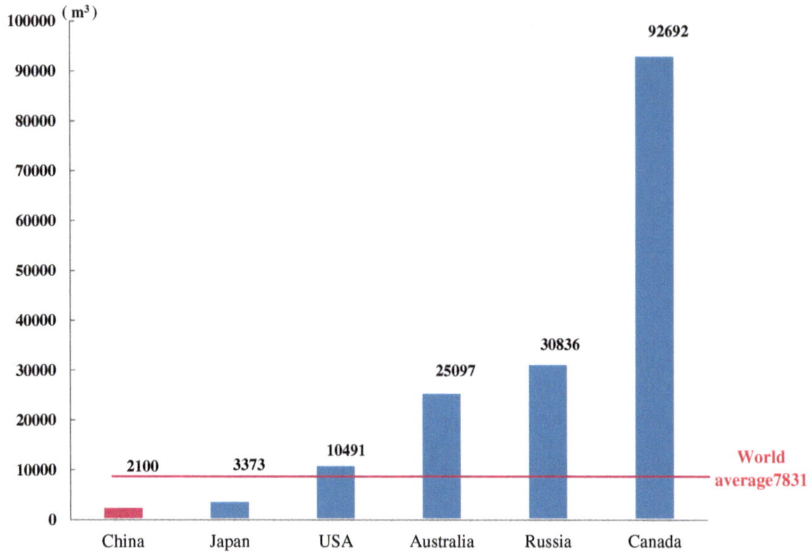

Fig. 1 China's water resources per capita compared to other countries

1.2 China Water Pollution Overview

According to the *Report on the State of the Ecology and Environment in China 2017*, out of 1,940 monitored surface water profile nationwide, 1,317 (67.9%) met Grade 1 and 3 water quality standards, while 462 (23.8%) reached Grade 4 and 5 standards. A total of 161 sections (8.3%) did not attain the Grade 5 standard (Ministry of Water Resources of the People's Republic of China, 2017).

1.2.1 Reservoirs

In 2017, among the 112 principle reservoirs surveyed, the water quality classifications were as follows: Grade I and II standard: 33 lakes (29.5%); Grade III standard: 37 lakes (33.0%); Grade IV standard: 30 lakes (19.6%); Grade V standard: 8 lakes (7.1%); Below Grade V standard: 12 lakes (10.7%). The primary pollution indicators were COD, TP, and the $KMnO_4$ index. In terms of nutritional status, out of 109 monitored lakes and reservoirs, 9 were oligotrophic, 67 were mesotrophic, 29 experienced slight eutrophication, and 4 were in an intermediate eutrophic state.

1.2.2 Rivers

In 2017, among the 1,617 monitored water sections across 7 major watersheds in China (The Changjiang River, Songhua River, the Zhujiang River, Yellow River,

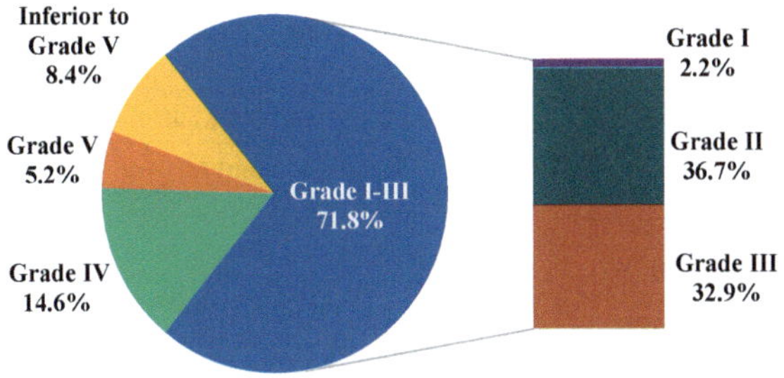

Fig. 2 Water quality grades distribution nationwide in 2017

Haihe River, Liaohe River, Huaihe River) and rivers in Fujian, Zhejiang, southwestern and northwestern regions, the water quality distribution was as follows: Grade 1: 35 sections (2.2%); Grade 2: 594 sections (36.7%); Grade 3: 532 sections (32.9%); Grade 4: 236 sections (14.6%); Grade 5: 84 sections (5.2%); Below Grade 5: 136 sections (8.4%) (Fig. 2). The primary pollution indicators were TP, ammonia nitrogen, and COD.

Overall, southwest and northwest China rivers maintained good water quality standards. The Changjiang River, the Zhujiang River, and rivers in Fujian and Zhejiang consistently met good quality standards. In contrast, the Huaihe River, the Huanghe River, Liaohe River, and Songhua River exhibited slight pollution, while the Haihe River faced moderate pollution.

1.2.3 Groundwater

In 2017, a total of 5,100 monitoring sites were established to assess groundwater quality. Among these sites, 8.8% were classified as excellent, 23.1% as good, 1.5% as relatively good, 51.8% as poor, and 14.8% as very poor in terms of water quality (Table 1). Key pollution indicators encompassed manganese, total dissolved solids, total hardness, iron, NO_2-N, NO_3-N, fluoride, NH_3-N, chloride, and sulfate.

1.2.4 Marine Water

In the summer of 2017, around 96% of the total marine area adhered to Grade I water quality standards (refer to Table 2 for specifics). The remaining portion, totaling 3,700 km^2, did not meet Grade IV standards.

Table 1 Groundwater quality assessment results across various catchments in 2017 (Guo et al. 2008)

River basin	The percentage of monitoring points (%)		
	Excellent, good and relatively good	Poor	Very poor
Huanghe River	26.8	45.7	27.5
Liaohe River	8.8	81.0	10.2
Haihe River	31.4	52.8	15.7
Songhua River	11.2	81.4	7.4
Inland River	39.1	47.8	13.0
Changjiang River	14.3	80.0	5.7
Huaihe River	24.4	67.3	8.2
Nationwide	24.4	60.9	14.6

Table 2 Marine areas not meeting Grade I standards (Guo et al. 2008)

Marine area	Season	Sea area with different water quality levels (km^2)			
		Grade 2	Grade 3	Grade 4	Inferior to Grade 4
Bohai Sea	Summer	8,940	3,970	2,120	3,710
	Autumn	15,710	8,300	4,780	3,690
South China Sea	Summer	6,000	8,220	2,110	6,560
	Autumn	11,900	8,900	4,210	5,270
Donghai Sea	Summer	17,610	9,260	11,400	22,210
	Autumn	23,280	10,260	11,850	34,510
Yellow Sea	Summer	17,280	7,090	2,610	1,240
	Autumn	20,980	10,980	9,440	3,840
All seas	Summer	49,830	28,540	18,240	33,720
	Autumn	71,970	38,440	30,280	47,310

2 Challenges and Prospects for Addressing Water Pollution

2.1 Eutrophication

2.1.1 The Status and Consequences of Lake and Reservoir Eutrophication in China

Eutrophication is a primary indicator of the natural aging process of lakes, linked to the accumulation rate and nutrients levels in lakes, making it the primary indicator. Nutrients in lakes originate from diverse sources like dust deposition, domestic sewage, precipitation, surface runoff, industrial wastewater, and sediment.

2.1.2 Hazards Stemming from Eutrophication in Lakes

(1) Water Quality Eutrophication, as seen in lakes, results in the rapid multiplication of cyanobacteria and green algae, causing a sharp increase in phytoplankton. This process elevates suspended solids levels and generates significant organic matter with a foul odor. This results in an increase in the water's nutrient content, resulting in the formation of "water blooms" that emit unpleasant odors and disrupt the water's function, impacting the surrounding environment upon decomposition.
(2) Eutrophication affects water supply and aquaculture by causing excessive algae growth in lakes and reservoirs. This growth clogs water plant filter tanks, reduces filtration efficiency, and increases treatment costs. For instance, Chaohu Fourth Water Plant experienced issues with strong-smelling water due to excessive algae. Fish and aquatic organisms may suffocate as a result. For instance, from the 1960s to the 1990s, fish production in Dongting Lake and Duyang Lake dropped by 30%. A significant fish die-off occurred in Dianchi Lake due to overgrown aquatic plants like water hyacinth, resulting in substantial economic losses in September 1986. Furthermore, eutrophication can lead to the disappearance of valuable fish species, significantly impacting aquaculture's economic benefits.
(3) Impact on Tourism Development Eutrophication in lakes and reservoirs causes algae proliferation, reducing water transparency and resulting in turbid, malodorous water bodies. These factors significantly diminish the appeal of lake tourism and sightseeing, sometimes rendering them ineffective.

2.1.3 Sources of Pollutants in Lakes and Reservoirs

Water pollution primarily results from human activities. The external sources of pollution in lakes and reservoirs can be categorized into two types:

1. Point source pollution originates from fixed discharge points, primarily stemming from industrial wastewater and domestic sewage discharged into lake basins through pipes.
2. Non-point source (NPS) pollution, in contrast, arises from numerous diffuse sources and can impact water bodies through various means, such as runoff from agricultural areas entering rivers. Urban runoff, which includes rainwater washing contaminants from parking lots, highways, roads, and lawns, is also considered NPS pollution. Suburban and urban areas are significant contributors to NPS contamination due to runoff produced by extensive ground leveling, facilitating transport of pollutants into nearby water bodies.

Under specific environment, external contamination can accumulate, resulting in endogenous contamination characterized by the continuous buildup of various pollutants.

2.1.4 Processing Technologies and Real-World Examples

Eutrophication can be addressed through physical, chemical, and biological treatments.

1. **Chemical Treatment**: This method involves the addition of chemicals and water absorbents to modify the pH and redox potential of polluted water while aiding in the adsorption of organic materials and suspended sediment.
2. **Physical Treatment**: Physical measures include artificial aeration, pollution containment, diverting water to flush out pollutants, and river dredging, all aimed at repairing polluted water.
3. **Biological Treatment**: Ecological restoration methods utilize technologies such as artificial floating islands, microbial agents, wetlands, biofilm, and biological grids. These approaches focus on leveraging natural processes and organisms to restore water quality.

Ecological Recovery Project in West Wuli River

The West Wuli Lake, located near Wuxi City, Jiangsu Province, China, is an extremely eutrophic shallow lake. A comprehensive ecological restoration experiment (10×104 m^2) was initiated in its southwestern region in January 2004. The primary measures included establishing a large enclosure to enhance submerged macrophyte growth, fish removal, sediment dredging, sediment rehabilitation, planting aquatic macrophytes, stocking piscivorous fish, and introducing benthic animals. Twenty-three aquatic macrophyte association comprising 15 families, 22 genera, and 32 species were established, leading to an increase in macrophyte coverage from 0 to 40 to 55%.

Inside the enclosure, water quality significantly improved, with concentrations of TP, TN, NO_2-N, Nitrate nitrogen, NH_3-N, and orthophosphate from August 2004 to July 2005 showing reductions of 23.8, 20.7, 45.6, 21.1, 35.2 and 54.0% compared to outside the enclosure. Water transparentness also increased signally to 0.70 m, from 0.39 m. These consequences emphasize the importance of restoring the ecological function and structure along the lakeshore, with biomanipulation serving as a valuable approach to reduce algal biomass, enhance water transparency, and ameliorate overall water property in blank areas, contributing to the entire lake's restoration.

Artificial Wetland

Constructed wetlands (CWs) provide cost-effective solutions for treating industrial, municipal, and agricultural wastewater. They are commonly used for nutrient removal from polluted streams. Case in point, wetland treatment systems can reduce TN and TP levels in hypereutrophic lake water by approximately 30–67% and 30–52%, respectively (Coveney et al. 2003). These CWs, known for their affordability, ease of operation, and maintenance, hold promise for addressing water pollution challenges in developing countries.

To address water pollution issues, especially in lakes like Taihu Lake, the implementation of new, cost-effective, and eco-friendly approaches is urgently needed. Constructed wetlands have the potential to effectively address nutrient pollution and contribute to the restoration of lake ecosystems. However, before taking concrete actions, it's essential to carefully design and run pilot-scale CW systems to evaluate their capabilities in purifying eutrophication waters. In Lake Taihu, South China, Li et al. conducted a 2004 study involving three parallel pilot-scale CW units, each covering 30 m^2, with different water-flow formats. The efficiency of constructed wetlands in removing pollutants can be influenced by factors like the wetland's aerobic and anaerobic environments and variations in water flow formats (Linfeng et al. 2008).

Compound Microorganisms

In a study conducted to explore techniques for controlling urban lake eutrophication in Lake Simeitang, Wuhan, *Bacillus subtilis* and *Rhodopseudomonas palustris* were introduced into the water from August 1 to September 27, 2008 (Ding-Xin et al. 2010). The findings revealed varying reductions in pollution indicators, including Chl-a, TN, TP, NH_3-N, and COD_{Mn}, with removal efficiencies reaching 46.70, 49.26, 73.02, 16.18 and 52.78%. Within one month of applying complex microorganisms, the visual state of the water intensely ameliorated, with the disappearance of unpleasant odors, water blooms, and increased transparentness. After 3 months, water feature remained stable. This cost-effective, rapid, and environmentally friendly technology, free from secondary contamination combined with extra eutrophication treatment methods, holds significant potential for widespread use in bioremediation of eutrophic loughs.

A study of South China Botanical Garden, Chinese Academy of Sciences, where frequent algae blooms occurred, examined the impact of effective microorganisms (EM) in controlling algae blooms in eutrophic water (Mei et al. 2000). Results showed reductions in Chl-a content at the water surface and 0.5 m below, as well as decreases in TN, TP, and CODMn after EM application. Simultaneously, dissolved oxygen (DO) content increased, along with improved water transparency. The substantial reduction in Chl-a indicated effective algae bloom control, directly contributing to increased water transparency. Decreases in TN, TP, and organic matter content appeared to be the primary factors responsible for algae reduction.

2.2 Challenges and Opportunities in Addressing Water Pollution

China's rapid economic growth has led to substantial growth in pollutants originating from various sources, including industry and agriculture. Due to limited environmental awareness, inadequate effluent discharge supervision, and insufficient environmental infrastructure, various pollutants, including inorganic compounds (e.g., NH^{4+}, NO^{3-}, PO_4^{3-}), organic substances (e.g., dyes, pesticides, and compounds from human daily life), and heavy metals (e.g., Hg^{2+}, Cr^{6+}, Cd^{2+}, and Pb^{2+}), are released into rivers. This leads to increased densenesses of phosphorus, N, COD, and other contaminants in water mass, exceeding their natural self-purification capacity. Consequently, the ecosystem balance within these water bodies is severely disrupted, leading to the formation of black and smelly water bodies (Xu et al. 2015; Ji et al. 2016). These water bodies often display foul odors, diminished aquatic life, and a notable decline in river ecosystem functioning (refer to Fig. 3). These issues present substantial challenges to city sustainability and can adversely affect health.

The problem of black and odor pollution in water bodies continued in 2018, according to the report of the "National Black and Odor Monitoring Platform for Water Bodies" established by the Ministry of Urban–Rural and Housing Development. Approximately 2,100 such water bodies were found in urban areas across China, affecting over 70% of cities. Significantly, economically developed provinces like Anhui, Jiangsu, Shandong, Guangdong, Hunan, and Hubei accounted for more than 71.9% of these polluted water bodies (Fig. 4).

To combat inkiness and stinking water, the Chinese governance has allocated significant funds and enforced environmental regulations. This has greatly improved water quality and spurred advancements in water cleaning technologies in affected areas. According to messages from the "National Monitoring Platform for Black and Odor Water" (http://gz.hcstzz.com/), around 1,745 urban black and smelly water

Fig. 3 The images depict the black-odorous Yawei rivulet in Haikou city, China, captured in 2016. These visuals are sourced from the Tenure of Bionomics Environment of China. (http://www.mee.gov.cn/home/ztbd/rdzl/hcszz/grb/201811/t20181112_673407.shtml)

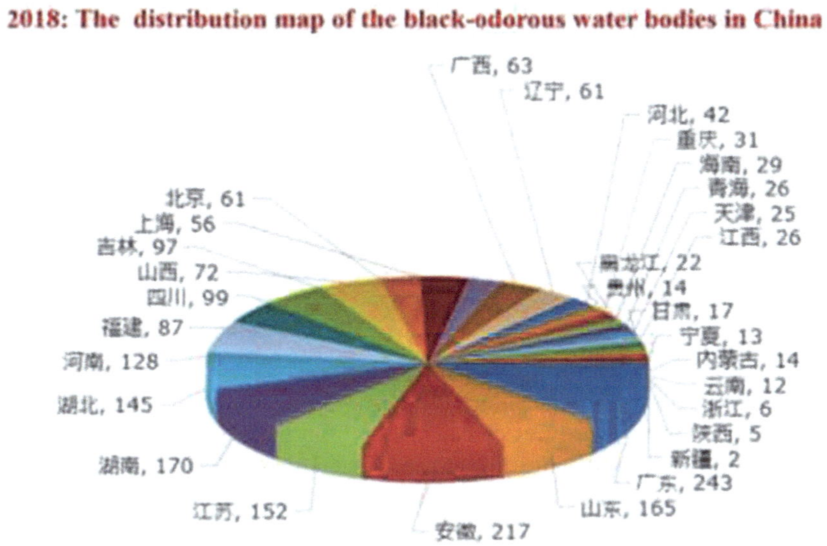

Fig. 4 The figure displays the distribution map of black and odorous water bodies in Chinese cities as of 2018. This graphic is sourced from the **Ministry of Urban–Rural and Housing Development of China** (http://gz.hcstzz.com/)

had received discourse, accounting for 83.1% of the whole affected water bodies. Currently, 264 are under treatment, and 91 are in the planning stages, accounting for 12.6% and 4.3%, respectively. It is projected that by the end of 2020, the proportion of inkiness and stinking water bodies in populated areas will be reduced to 10%, with the goal of complete elimination by 2030.

However, the challenge of addressing inkiness and stinking water in China, especially in countryside (Currell et al. 2010), remains significant. The water environment ecological remedy, aiming at eliminating black and stinking water, will continue to be a crucial aspect of China's ongoing efforts to control water pollution (Wang et al. 2018; Zhaohai et al. 2018).

2.3 *Groundwater*

China's rapid social and economic development has resulted in significant discharges of domestic and industrial wastewater into surface waters, leading to pollutant accumulation in lakes, rivers, and reservoirs (Ito 2005) (Fig. 5). These pollutants are gradually seeping into groundwater, causing a notable decline in its quality (Rauret et al. 2012). This issue has gained widespread attention in Chinese society. Currently, human activities are increasingly pressuring groundwater resources (Bo et al. 2013; Li et al. 2013; Sun et al. 2015). In certain regions, declining groundwater levels are

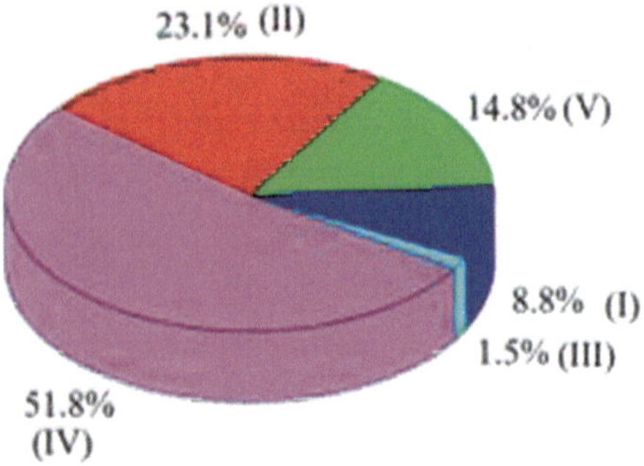

Fig. 5 Groundwater quality grade in China (Ministry of Ecological Environment of China 2017)

negatively affecting both its quantity and quality, as well as ecosystems reliant on it. Furthermore, groundwater faces contamination from both non-point source pollutants, like those from agriculture (Trauth and Xanthopoulos 1997), and point source pollutants, originating from industrial activities (Frische et al. 2010).

Groundwater quality is shaped by both natural and anthropogenic factors. Natural factors involve aquifer lithology, water replenishment quality, and interactions between groundwater and rocks. Anthropogenic factors encompass activities like agriculture, industry, water resource extraction, and urban development (Currell et al. 2010; Nazzal et al. 2014; Rouabhia et al. 2010). Consequently, the primary contaminants in groundwater fall into three main groups: heavy-metals, nutrients, and organic contaminations (Table 3).

In recent years, the increasing severity of groundwater pollution has drawn significant attention. Consequently, the application of various methods, such as physical, chemical, or biological approaches, for remediating contaminated groundwater has gained widespread interest. These groundwater contamination remedy technologies can be categorized into 2 primary groups based on their remediation approach: in-situ and ex-situ remedy methods.

Table 3 Quality assessment of groundwater in various basins across China (Ministry of Ecological Environment of China 2017)

Basin	1~3 (%)	4 (%)	5 (%)
Huaihe River	24.4	67.3	8.2
Changjiang River	14.3	80	5.7
Haihe River	31.4	52.8	15.7
Huanghe River	26.8	45.7	27.5
Songhua River	11.2	81.4	7.4
Liaohe River	8.8	81	10.2
Inland River	39.1	47.8	13
Whole	24.4	60.9	14.6

2.4 Sea

Sea pollutants in China result from a combination of human and natural influences. These pollutants can be categorized into four primary types based on their specific sources:

1. Land source pollution, arising from industrial waste, household garbage, and agricultural runoff due to human activities, is a significant contributor to marine pollution. Notably, over 80% of marine pollution in China results from land-based pollutants (Fig. 6).
2. Marine industrial pollution, typically arising unexpectedly during marine oil extraction, marine accidents, shipping operations, and related factors.
3. Pollution from offshore farming and fishing mainly results from the unsustainable enlargement of aquaculture and the excessive victimization of pelagic resources.
4. Air contamination refers to the introduction of contaminations into the pelagic environment through rainfall and air movement (Jin et al. 2009).

2.5 Policy

2.5.1 Implementation of Water Laws, Policies, and Constitutions of China Pertaining to Water Contamination Control

China water contamination has caused significant environmental damage, with the Guanting reservoir being a prime example. This pollution not only affected fish populations but also resulted in people contracting Minamata disease due to consuming contaminated fish from polluted rivers (Zhou 2009). These incidents prompted Chinese leaders to recognize the critical need for environmental protection, particularly when the government prioritized drinking water safety in response to severe pollution in both the Tai River and Huai Lake. China's water law system comprises extensive legislation, including four primary laws, nineteen managerial regulations, fifty-five ministerial rules, and approximately seven hundred sub-national government rules and regulations (Ministry of Water Resources, The People's Republic of China 2017a).

Environmental Impact Assessment System (EIA)

Established in 2003, this system requires all project promoters to submit a report outlining the evaluation of proposed prevention measures and potential adverse environmental impacts to the relevant local authorities. Its implementation has had a substantial impact.

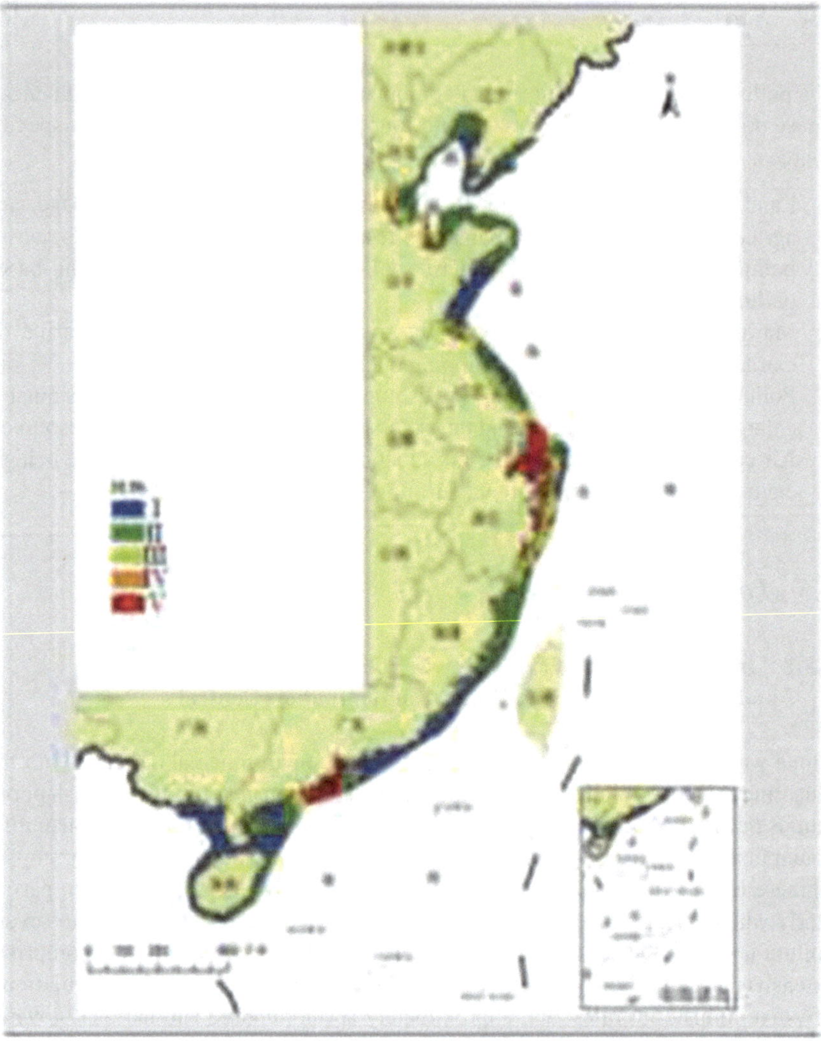

Fig. 6 Offshore pollution in China (Ministry of Ecology and Environment of the People's Republic of 2017)

The "Three Simultaneous System"

The "Three Simultaneity system," which requires projects with environmental hazards to incorporate and utilize pollution control equipment during operation, is closely related to the previous EIA system. While the EIA system concentrates on devising water pollution prevention schemes, the "Three Simultaneity System" emphasizes their practical implementation. Notably, this system has made a positive impact on controlling pollution from numerous newly established industrial point pollution sources, and its implementation trend remains consistently high.

Pollution Levy System

The pollution levy system grants local governments the authority to impose charges on entities, particularly enterprises, that exceed emission standards. This system comprises two components: the first one goals to penalize pollutant discharge through pollution levies, while the second addresses issues related to resource occupation and waste discharge (Zhou 2009).

Emission Reporting System

Under this scheme, enterprises are required to furnish local environmental authorities with a comprehensive list of pollutants, including their quality, quantity, and concentration, along with proposed measures for reduction. This approach facilitates the creation of a comprehensive local pollutant database, enabling local governments to effectively strategize environmental protection efforts.

2.5.2 Dares and Chances of Water Policy to Control Water Contamination in China

Challenges

China's environmental legislation faces some challenges:

(a) There is a need to fully integrate maintainable advancement into environmental laws. Additionally, there is a requirement to emphasize the principles of "prevention in the first place" and "Prevention priority, combining prevention with control" rather than relying on "treatment in the end."
(b) Insufficient coordination between laws and regulations persists, often due to incomplete relevant regulations, hindering effective implementation.
(c) Challenges include issues with responsibility assignment, system design, disparities in rights, and a lack of operability within the legal framework.

Chances Arising from Chinese Water Policies

The evolution of China water law primarily focuses on environmental protection and may involve stricter measures. However, it's worth noting that water control regulations in China could also stimulate various developments:

(a) Promoting investment in acquiring advanced foreign water pollution control technology.

(b) Promoting the establishment of public–private partnerships to delegate water treatment to specialized third parties, as highlighted by van de Bovenkamp (2016).
(c) Facilitating the growth of local businesses specializing in wastewater monitoring and wastewater treatment services.
(d) Potentially leading to an increase in employment rates and related economic benefits.

3 Addressing Water Pollution in a Rapidly Changing Environment: Insights from China

3.1 Origin Direct and Pollutant Interception

Origin control and pollution interception form the fundamental prerequisites for addressing water pollution (Qiqi et al. 2013). Source control involves diminishing pollutant entry into water bodies at their source, while pollution interception focuses on reducing pollutant entry during transport. These methods, while highly efficient, often necessitate extensive projects and significant investments. China has developed various strategies and technologies, and some of these will be discussed below.

3.2 Point Origin Pollution Direct

Point source contamination is characterized by an identifiable, single source. Unlike nonpoint sources, point sources are characterized by their limited spatial extent, often approximated as mathematical points to facilitate mathematical modeling. Pollution point sources encompass facilities like municipal sewage treatment plants, power plants, factories, and industrial stormwater discharge pipes.

Several technologies and methods have been advanced to mitigate point-origin contamination, including upgrades to wastewater effluents, among other approaches.

3.3 Non-Point Origin Direct

Non-point source contamination, often referred to as the entry of pollutants into water bodies through extensive, random, and accidental means, lacks fixed discharge outlets for these non-point pollutants. Urban non-point source pollution primarily includes livestock wastewater, initial rainwater, and solid waste. The main features of this type of pollution in China include:

1. Severe pollution levels.
2. Predominantly consisting of COD, SS, and nutrients.
3. Exhibiting regional and temporal variations.

Currently, widely employed engineering measures for urban non-point source pollution control include rainwater collection and treatment systems, as well as initiatives like sponge cities.

3.4 Sponge City

The term "sponge city" represents an innovative concept. This novel framework for water ecological management not only enhances urban greenery but also harnesses rainwater collection to naturally safeguard the city's water systems (Xia et al. 2017a, b). It's important to note that "sponge" here isn't an abstract idea; it corresponds to tangible landscape patterns. Building a sponge city involves creating the necessary water ecological infrastructure, which happens to be the most efficient and resource-intensive way of utilizing water resources. Consequently, the construction of sponge cities closely ties into the landscape design field.

Traditional town development often disrupts the natural flow of rainwater. In contrast, sponge city construction relies on a series of engineering techniques and landscapes to emulate the natural processes of rainwater absorption, storage, and evaporation. This approach transforms the urban drainage system to align with the natural rainwater circulation pattern (see Fig. 7).

Sponge city construction encompasses several key features:

1. **Infiltration**: By minimizing hard pavement, it maximizes the use of natural surfaces for infiltration, reducing runoff, preserving the ecology, and accumulating water resources.
2. **Stagnation**: Utilizing rainwater retention systems enhances stagnation and infiltration, slowing down the pace of rainwater collection, lowering drainage intensity, and reducing disaster risk.
3. **Storage**: It involves adjusting the spatial and temporal distribution of rainwater and creating conditions for its utilization.
4. **Net**: Reduction of non-point source pollution, lowering quantities of major pollutants like total nitrogen, suspended matter, chemical oxygen demand, and total phosphorus, thereby improving urban water quality.
5. **Use**: Emphasizes efficient utilization of rainwater resources and reclaimed water to enhance water use efficiency and alleviate water scarcity.
6. **Drainage**: Establishing storage and drainage systems to prevent waterlogging and ensure urban operational safety.

In essence, a sponge city represents a new generation of urban stormwater management. It operates by absorbing, storing, seeping, and purifying water, releasing stored water as needed. The "sponge bodies" within a sponge city encompass not only

Fig. 7 The illustration for sponge city. This figure is from the network station: http://blog.sina.com.cn/s/blog_15074176b0102wv8y.html

natural water systems like rivers and ponds but also urban features such as green spaces, gardens, and permeable pavements. Rainwater infiltrates, purifies, and is repurposed through these cavernous bodies with any excess runoff ultimately directed into the pipe network and pump stations. This approach significantly improves the standards of the city guttering system and reduces the risk of waterlogging.

In the Zhenzhu Bay, one of Guangzhou city's cavernous bodies pilot regions, municipal road construction involves the use of penetrable path materials, for instance, penetrable and concrete bricks. Rainwater seeps through the pavement structure, enters the pipeline, and is conveyed to natural water treatment plants. After purification in these facilities, the water serves various purposes, including toilet flushing, car washing, and road maintenance. Furthermore, the incorporation of municipal green spaces, artificial wetlands, lakes, permeable pavements, and dished ditches effectively stores rainwater and reduces surface runoff following rainfall. The "sponge road" system in the Zhenzhu Bay forms a comprehensive penetrable network capable of intercepting more than 80% rain in normal weather.. This not only recharges groundwater and enhances water quality but also improves the local microclimate and mitigates the urban heat island consequence.

3.5 Nutrients in Sediments

Inner nutrient discharge of deposit is a primary contributor to water body pollution (Qin 2006). Pollutants diffuse from sediment into the water through the water-deposit interface due to natural and man-made procedures, resulting in continuous water contamination. Consequently, effective pollutant removal from deposit is essential for preserving excellent water quality in shallow, huge loughs. Typical contaminations found encompass carbohydrates, heavy metals, nutrient salts, and emerging contaminants.

Carbohydrates primarily consist of discharged organic matter (Fan et al. 2016), along with natural organic matter like humic and non-humic substances (Zhang et al. 2010). During severe water pollution stages, organic matter in sediment is often concentrated from the water by adsorption effects. Plant and animal residues also contribute significantly to organic matter content. Nutrient salts refer to the accumulation of N and P in sediment (Ni and Wang 2015). Nitrogen, a major pollutant, consists of nitrate nitrogen, ammonia nitrogen, and nitrite nitrogen (Chen et al. 2014). It primarily originates from both non-point and point-origin contamination in shallow, huge loughs. Organic NH_3, including proteides and amidogen acids, also contributes to the generation of nitrate nitrogen, ammonia nitrogen, and nitrite nitrogen through biogeochemical cycling to maintain a balance in sediment nitrogen content.

Heavy metal sediment pollution in large, shallow lakes has been observed in China (Chen et al. 1989), with a particular emphasis on common heavy metals such as Cd, Cr, Pb, Zn, As, Hg, Cu, and more. However, often overlooked metal pollutants such as Mo, Co, V, Ti, Sr, Tl, Sb, Mg, Ca, Be, and Li also pose threats to water safety. Heavy metals are deposited in sediment through adsorption (Wen et al. 1998), complexation (Wen et al. 1998), and precipitation processes (Fang et al. 2011), maintaining a dynamic equilibrium distribution between the sediment and water.

The rapid growth of the economy has led to the detection of numerous emerging contaminants in sediment, including persistent organic pollutants (POPs) (Minh et al. 2007), antibiotics (Yang et al. 2010a, b), microplastics (Cauwenberghe et al. 2015), and more. These emerging contaminants display significant toxicity, bioaccumulation potential, and resistance to biodegradation, which pose threats to both the human health and ecological environment. In summary, comprehending the conditions of internal nutrient release in the sediment of shallow, huge lakes represents the initial step towards restoring river sediment.

3.6 Ecological Restoration

Ecological restoration involves the restoration of the functions of damaged ecosystems, encompassing physical, chemical, and biological aspects (Deqi 2005).

For eutrophication of lakes and reservoirs, restoration techniques span physical, biological, and chemical approaches.

Physical approaches encompass artificial, sediment dredging, aeration, and mechanical salvage.

Chemical approaches encompass deactivation, acid–base neutralization, chemical precipitation, and chemical seaweed control (Zhen et al. 2011).

Biological methods involve ecological floating islands, biofilm technology (Zhen et al. 2011), phytoremediation (Zhen et al. 2011; Zou et al. 2013), aquatic animal remediation (Naumann et al. 2015), and wetland restoration, among others.

3.6.1 Physical Methods

(1) **Deposit Dredging**

Deposit dredging utilizes automatic means to remove polluted deposit (Xia et al. 2017a, b). This method effectively eliminates polluted deposit from reservoirs, improves the bottom surrounding, and eliminates algae accumulation at the sediment–water interface. Furthermore, it aids in rehabilitating aquatic ecosystems, mitigating algae blooms, and improving water quality. Sediment dredging has also been associated with increased community diversity (Ning et al. 2004; Fischer and Dietrich 2000; Zhong 2009).

Sediment poses a potential threat to both humans and the aquatic ecosystem when pollutant concentrations are 2–3 times higher than background levels, necessitating a reevaluation of the need for sediment dredging.

Nonetheless, sediment dredging faces several challenges:

1. **High Financial Costs**: Dredging operations require substantial financial resources. For instance, dredging in Dianchi Lake involved the excavation of 4 million tons of sediment, 250 million yuan (Huiguang and Pijing 2000).
2. **Ecosystem Disruption**: Deep dredging can disrupt the original ecosystem.
3. **Complex Sediment Treatment**: Sediment's complex pollutant composition and high-water content make treatment challenging.

Environmental protection dredging aims to enhance water quality and restore ecosystems. Through the use of physical methods, this engineering effort reduces nutrient, toxic organism, heavy metal, and other pollutant levels in sediment. It also minimizes sediment-related ecological risks and facilitates safe sediment treatment and disposal (see Fig. 8) (Xia et al. 2017a, b).

Based on both domestic and international research, sediment dredging technology has shown clear effectiveness in smaller water bodies. However, in larger water bodies, particularly lakes of medium to larger sizes, there hasn't been a successful case. This suggests that the effectiveness of this method is limited (Deqi 2005).

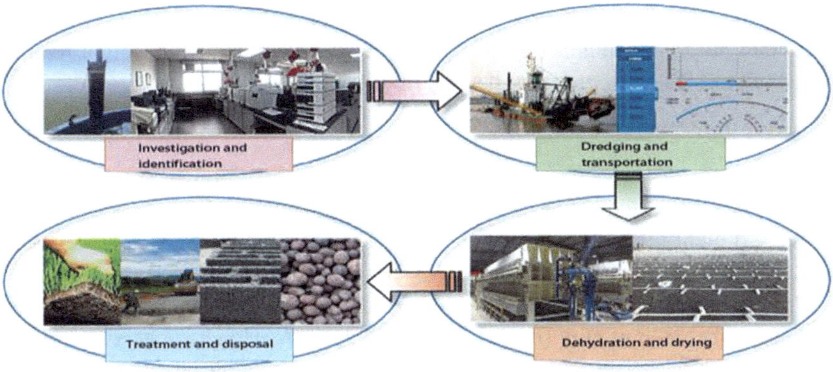

Fig. 8 Environmental protection dredging engineering

(2) **Contamination Control Engineering**

Contamination control engineering proves effectual in managing exogenic nutrients in eutrophic lakes. However, its impact on these lakes isn't particularly significant. For instance, in Xuanwu Lake, the total exogenous nutrient load has increased over decades, yet the whole consistence of P and N in the lake has only marginally decreased. Even though we could entirely intercept exogenous pollutants, a substantial nutrient reservoir would still exist within the lake, and eutrophication wouldn't be alleviated. Furthermore, some exogenous pollutants are impossible to completely eliminate. Simulation experiments conducted in West Lake demonstrated an increase in Chl-a concentration and algal blooms after pollution control measures, whereas the control group did not experience a bloom (GuoXiang and Peimin 1999). In summary, effectively managing algal blooms in eutrophic lakes is challenging using a single decontamination method.

(3) **Mixing Sewage with Water**

Diluting sewage with water can be effective in lakes with high nutrient and algae concentrations, reducing nutrient levels and algal density in eutrophic water. However, the dilution capacity of this method is influenced by various factors, including water quality and quantity.

Research conducted on cyanobacteria in different West Lake areas demonstrated that discharging draining sewage into Qiantang River led to a decrease in cyanobacteria from 84.4 to 69.7% and an increase in species diversity from 118 to 192. Importantly, this discharge did not significantly alter water quality. Interestingly, algae density tended to rise. Notably, when the discharge of diluted sewage ceased for a few days, the lake's total phosphorus (TP) concentration increased once more (Meizi et al. 1992).

However, for lakes afflicted by severe endogenous pollution, diluting sewage with water may not yield satisfactory results. Controlling eutrophication often requires more comprehensive approaches beyond dilution alone.

(4) **Aeration**

Density stratification is common in deep lakes and reservoirs, where oxygen levels in the lower waters decrease due to limited oxygen supply. In anaerobic environments, sediment releases both organic and inorganic contaminants, contributing to the deterioration of the aquatic ecosystem (Singer et al. 2008).

Oxygenation and mixing through compressed air represent a clean in-situ method for improving water quality (Fabbricino and Petta 2007; Gómez et al. 2000; Huang et al. 2006, 2008; Rivas et al. 2000). This method avoids introducing side effects or disrupting the biological environment and aquatic chemical balance, avoiding additional physical, chemical, or biological complications. In-situ oxygenation and water quality control mixing technologies encompass rising air bubbles, deep aeration-oxygenation, and water lifting using water jets (Singer et al. 2008).

Deep aeration, adds oxygen to the lower water column, increasing dissolved oxygen levels and inhibiting pollutant release from sediments (Sheffield et al. 1989). However, due to limited water circulation, deep aeration has limitations in dispersing oxygenated water to surrounding areas and cannot directly control algae growth.

Bubble plume mixing employs diffusers at the lake bottom to release air which enters the water in the form of small bubbles. As these bubbles rise, they entrain lower water, carry it upwards in a plume, spread it across the surface area, and then descend, creating vertical circulation. This circulation disrupts thermal stratification and facilitates oxygen transfer into the water. Additionally, it transports surface algae to lower and middle waters beyond the bubble plume, where reduced light intensity and lower temperatures inhibit algal growth. However, the mixing zone has limited range, and air tube installation requirements are relatively high, making it challenging to implement (Huang 2016).

More recently, premium water aerate technology has emerged to develop in-situ water quality and combat eutrophication in reservoirs (Huang 2016). Water-lifting aerators (WLAs) combine features of air-lifting aerators and hypolimnion aeration, effectively mixing upper and lower water layers and oxygenating the water (Huang 2016).

3.6.2 Chemical Methods

Chemical restoration techniques are employed to either control algae by managing nutrient concentrations in water or eliminate algae using chemical agents (Deqi 2005).

1. **Controlling Nutrient Concentrations in Water**: This approach reduces nutrient levels in water through chemical processes. To mitigate phosphorus (P) concentration and combat eutrophication, ferric salts or aluminum salts are introduced to

create chemical precipitation of inorganic phosphates via flocculation or adsorption in the water. Aluminum salts can be precipitated into aluminum hydroxide, which adheres to sediment surfaces, preventing the discharge of deposit P. The addition of lime helps regulate water acidity and alkalinity, promoting to form stabilized $Ca_{10}(PO_4)_6(OH)_2$ precipitation and control Chl-a and phosphorus concentrations in the water.

2. **Algae Elimination with Chemicals**: This method employs chemical agents to exterminate algae. Current algaecides, synthesized and selected for this purpose, fall into two categories: organic compounds such as organic acids, aldehydes, ketones, and quaternary ammonium compounds, and inorganic substances like potassium permanganate and copper salts. While chemicals demonstrate significant efficacy in algae control, prolonged chemical use can have adverse effects.

3.6.3 Biological Methods

(1) **Ecological Floating Island**

Ecological floating islands involve planting vegetation on floating platforms. These plants and the microorganisms that attach to their roots are utilized to reduce water pollutants (Hongmei et al. 2017).

The ecological floating bed serves various functions, including 1. Provide habitats for fish and birds. 2. Absorb nitrogen and phosphorus while purifying water quality. 3. Enhance the aesthetic appeal of the environment. 4. Mitigate waves and safeguard lake shores (Yong and Congjun, 2009).

While ecological floating islands have proven effective in water restoration, they have some limitations (Shumei, 2016): 1. Lack of technical standards for plant and carrier selection. 2. Effective treatment but lengthy processing times. 3. Challenges associated with disposing of plant material.

Zhou and Yang et al. conducted experiments using ecological floating islands planted with yellow flag plants. Results indicated that these floating beds significantly reduced CODMn, TN, and TP. During the static stage, the net removal rates for CODMn, TP, and TN reached 16.9%, 50.0%, and 50.3%, respectively. Removal rates decreased as hydraulic retention time reduced. But even with the shortest hydraulic retention time (4.5 days), net removal rates for CODMn, TN, and TP remained at 9.4%, 39.8%, and 36.6%, meeting the national surface water quality requirements of the third level. This demonstrates that ecological floating islands can significantly improve eutrophication through purification efforts (Yang et al. 2010b).

(2) **Ecological Revetment**

Ecological revetment is a method that combines plants with civil engineering to safeguard lake slopes (Xubao 2013).

The ecological bank protection functions encompass:

1. **Water Storage and Runoff Retention**: During the wet season, revetments can store water and recharge groundwater. The plants on these revetments also help retain runoff. During the dry season, groundwater can be returned via revetment.
2. **Erosion Prevention**: Ecological bank protection employs plants with robust root systems, effectively preventing soil and water erosion and enhancing the revetment's resistance to erosion.
3. **Habitat Creation**: The porous nature of the revetment's slope foot provides a habitat for animals, amphibious animals, and aquatic plants.
4. **Environmental Enhancement**: Ecological shoreline conservancy creates a lovely natural landscape, and protects the ecological environment.

Ecological bank protection can be categorized based on the revetment materials into two main types: Natural Ecological Shore Preservation includes pine log shore protection, plant shore protection, and riprap shore protection; Artificial Ecological Bank Preservation encompasses ecological concrete shore protection, concrete lattice shore protection, gabion shore protection, concrete block shore protection, geotechnical material shore protection, eco-bag shore protection, and porous brick shore protection (Xiaopeng 2017) (Fig. 9).

(3) **The Biomembrane Technology**

Biomembrane technology employs natural or synthetic materials as carriers, which offer a large specific surface area for microorganisms to attach (Hang et al. 2010). These carriers effectively adsorb, intercept and degrade pollutants in sewage, while also converting excess nutrients like N and P for water purification (Zhen et al. 2011).

The biofilm process consists of four stages: (1) Contaminants move to the horizontal plane of the biofilm (External diffusion stage); (2) Contaminants move within the biofilm (Internal proliferate stage); (3) Catalysts and enzymes are secreted by microorganisms (Chemical reaction stage); (4) Metabolins are expelled (Release stage).

Biofilm technology offers benefits such as bacterial immobilization, adaptability to environmental changes, rapid pollutant degradation, and reduced sludge production. However, it comes with the drawback of high initial and operational costs.

Liuxuan et al. utilized a bamboo biofilm reactor for low-pollution landscape water treatment. Results demonstrated removal rates of 19.99–100% for turbidity, 15.9–59% for NH_3-N, and 10.60–74.34% for CODMn in water.

(4) **Wetland Restoration**

Wetlands encompass diverse chemical, physical, and biotic processes, leading to the detention, transformation, release, or production of various substances. Their purification capabilities arise from numerous distinct mechanisms. Wetlands are categorized based on topography into estuarine, embankment, lakeside, and steep bank types, each defined by UNESCO's MAB as transitional ecosystems distinct from water and land.

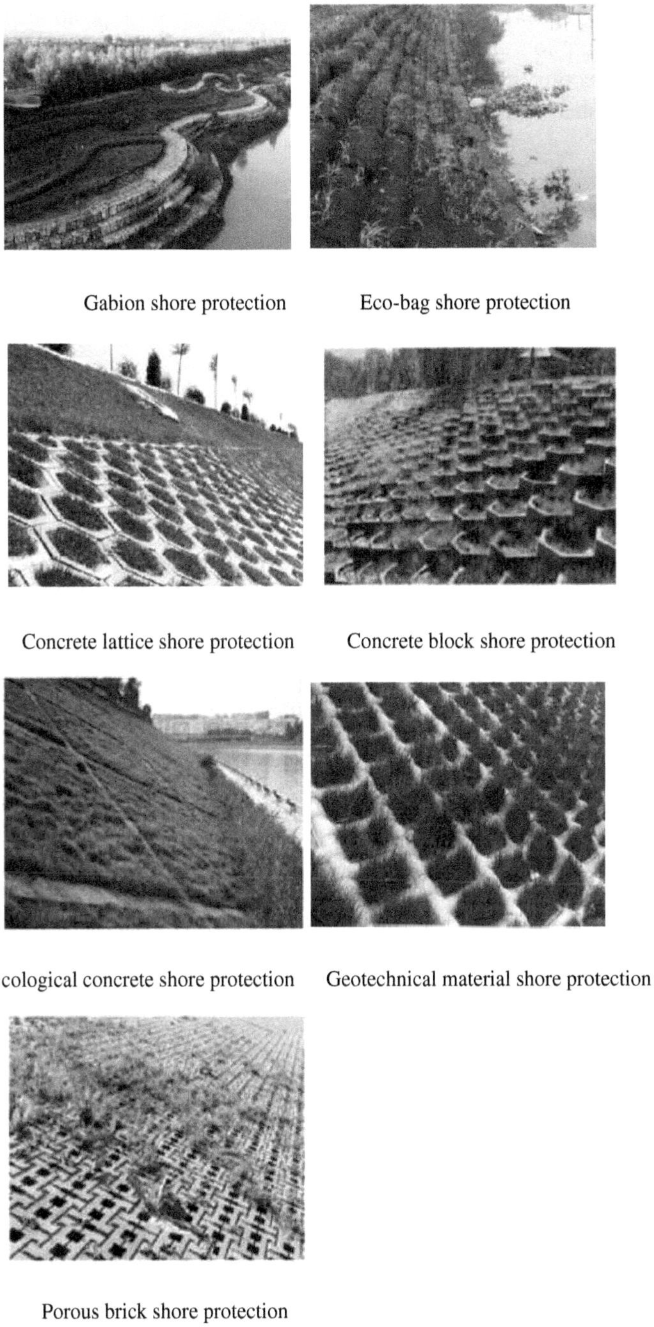

Fig. 9 Ecological shore preservation

Lakeside wetlands primarily purify through microbial decomposition, plant interception, soil adsorption, and plant absorption. These processes involve ion exchange, decomposition, complexation, adsorption, filtration, nitrification, precipitation, denitrification, and plant nutrient uptake (Zilu 2011).

Lakeside wetland ecosystems are vulnerable and easily disrupted. The Lakeside Wetland Park plays a crucial role in their restoration and reconstruction, contributing significantly to wetland ecosystem protection.

3.6.4 Restoration by Aquatic Animals

Aquatic animal-based restoration technology, a form of biomanipulation, aims to enhance water quality and restore aquatic ecosystem balance. Biomanipulation involves altering the biotic community structure, following ecosystem food chain principles and organism relationships.

Current biomanipulation measures include 1. Modifying zooplankton predator composition, species, and quantity to regulate zooplankton community. 2. Advancing the filter-feeding zooplankton population growth. 3. Decreasing planktonic algae biomass. 4. Enhancing water quality, often by breeding zooplankton to control algae.

Some scholars suggest that filter-feeding fish, like big head fish, can directly consume phytoplankton. Xie and Liu (2014) introduced non-traditional biomanipulation, controlling fish communities with stable structures to decrease planktonic algae through their unique digestive and feeding mechanisms. This method directly manages algae blooms. Introducing filter-feeding zooplankton (*Daphnia magna and Gerzoda daphnia gamaranaura*) yielded positive results in treating nutrient-rich lakes in Nanhu Lake.

Simulated experiments demonstrated the impact of *Hypophthalmichthys molitix* and *Aristichthys nobilis* carp ratios on water quality (Tong et al. 2016). A 4:1 ratio achieved a 73.65% NO_3-N removal rate, while a 5:1 ratio reached 66.73% removal rates for TP and PO_4^{3-}. Plagiognathops microlepis bleeker also effectively removed TP and PO43%, increasing removal rates by 13.89% and 7.91%, respectively. In Moon Lake, addressing cyanobacterial blooms through alum treatment and introducing *Hypophthalmichthys molitix* and *Aristichthys nobilis* prevented further blooms and improved water quality.

3.6.5 Phytoremediation

Phytoremediation is an economically viable alternative to costly cleaning methods. This approach utilizes floras to stabilize, divert, remove, or pollutant degradation in sediment, ground, and water, offering the following abilities.

(1) **Heavy-Metal Biosorption**

Hydrophytes are recognized for their ability to accumulate and concentrate heavy metals in ecosystems (Xingyu and Kefu 2001). Numerous studies confirm their effectiveness in shifting heavy-metals from contaminated water. For instance, Pistia stratoites utilizes its extensive root system for heavy metal removal, while water hyacinth can transform and absorb Cu, Cd, Zn, Pb, and Ni within its tissues. L. minor, a common aquatic plant, excels at removing up to ninety percent of soluble plumbum from water, thriving in the potential of Hydrogen levels ranging from six to nine. The aquatic plants capacity to absorb heavy metals varies by species and the type of metal (Singh et al. 2012). Additionally, their performance can be influenced by climate, geography, temperature, and environmental conditions.

(2) **Sanitized Water Body**

Research has demonstrated that hydrophytes play a dual role by restraining literary elegance growth and absorbing nitrogen and phosphorus to regulate water nutrient levels (Runan et al. 2011). In a study simulating Lake Taihu's water quality, ten aquatic plant species were cultivated in a laboratory setting, including *Hydrillaverticillata (Linn. f) Royle*, *Typha orientalis*, *Eichhornia crassipes*, *Lemna minor*, *Phragmites perfoliatum*, *Pistia stratiote*, *Potamogeton malaianus Miq*, *Zizania latifolia*, *Vallisneria natans L*, and *Ceratophyllum demersum Linn* (Yan 2013). Their findings revealed that various aquatic plants exhibit distinct rates of total nitrogen (TN) removal in eutrophic water. *Phragmites perfoliatum*, *Zizania latifolia*, and *Typha orientalis* demonstrated removal rates of 92.2%, 89.75%, and 87.64% for TN, respectively. Among them, *Phragmites perfoliatum* excelled in TN removal, while *Vallisneria natans* proved most effective at phosphorus removal (TP). Consequently, planting aquatic vegetation emerges as an efficacious strategy to combat loch eutrophication.

(3) **Allelopathy of Floras**

Allelopathy refers to the process where plants are influenced by discharging chemicals to the external to grow other plants, including microorganisms, often leading to inhibition (Rice 1984). Currently, various aquatic plants are known to produce substances, termed allelochemicals, that can effectively inhibit algae growth (Rouzic et al. 2016; Xiang et al. 2013). The utilization of allelochemicals is considered ecologically safe and is a viable approach for controlling harmful algae blooms (HABs) in natural lakes (Xiang et al. 2013).

(4) **Landscaping**

Apart from their water quality improvement functions, aquatic plants also enhance the environment's aesthetics. In practical applications, their ornamental value plays a significant role. Aquatic plants with varying leaf shapes, colors, and flower forms compensate for each other's strengths and weaknesses in terms of aesthetics. Consequently, a diverse selection of aquatic plants is employed to enhance their ornamental appeal.

3.7 *Water Transfer*

Water transfer involves relocating significant volumes of water between regions. This can serve multiple purposes, such as diluting pollutants in polluted rivers or lakes with clean water and addressing water shortages in water-deficient areas. Eutrophication, a major water issue in China, results in the gradual increase in biological activity in freshwater bodies, potentially leading to their extinction. Notably, some Chinese water bodies like Lake Taihu and Lake Baiyang suffer from severe eutrophication.

Engineering water transfer projects alter ecological environments by moving surplus water from one body to another. These projects have been successfully implemented worldwide. For example, the Snowy Mountain Project in southeast Australia transfers water over a 740 m gap among Sydney, Canberra, and Melbourne (Pigram 2000).

The California State Water Project transfers water from northern California to Southern California to address water needs in drought-stricken areas. In China, notable projects include the Great Canal water project, spanning 1794 km from Hangzhou to Beijing, initially designed for navigation (Yao 1998). Another prominent Chinese initiative is the South-to-North Water Diversion (Liu and Zheng 2002). Below is a table summarizing major water transfer projects in China.

3.7.1 Table Showing the Summary of Water Transfer Projects in China

Name	Country	Goal	Detailed description	Advantages/effects	References
South-North water transfer/diversion	China	Resolve water lack matters in the north of China	Supplying the northern region of China with about 44.8 billion cubic metres of water annually	Improved the irrigation, water supply, hydropower, flood control, water quality and ecology of the environment	Water Technology; Bloomberg News (2017); The Economist (2018)
Eastern route of water diversion project	China	Supply water to Shandong Province and the northern part of Jiangsu province		Improved the water supply, irrigation, hydropower, floodwater command, water quality and ecology of the environment	Water Technology; Bloomberg News (2017); The Economist (2018)

(continued)

(continued)

Name	Country	Goal	Detailed description	Advantages/effects	References
Western route of water diversion	China	Supply northwest China with water	4 billion cubic metres of water	improved the water supply, hydropower, floodwater control, irrigation, water quality and ecology of the environment	Water Technology; Bloomberg News (2017); The Economist (2018)
Great Canal water project	Australia	Moves water from Beijing, to Hangzhou	This is not only one of the longest (1794 km). but also the first manually operated canal mainly for navigation	improved the flood control, irrigation, water supply, hydropower, water quality and ecology of the environment	Water Technology; Bloomberg News (2017); The Economist (2018)

3.7.2 Importance of Water Transfer Within China

These water transfer initiatives have significantly enhanced irrigation, water supply, hydropower generation, flood control, and overall environmental quality. Notably, they have been effectively utilized for short-term solutions, involving the dilution of polluted water with freshwater, addressing eutrophication issues in loughs, for example, Moses Lake and Green Lake in the United States of America, as well as Lake Taihu (Hua et al. 2008).

Additionally, China South-to-North Water Diversion project exemplifies an ambitious strategy to combat water scarcity in the northern region by transferring water from the southern region (Liu and Zheng 2002).

3.7.3 Case Study

The Changjiang River-Taihu Lake Water Transfer Project

The water transfer project for Lake Taihu commenced in 2002. As the third largest freshwater lake, Lake Taihu has been severely affected by eutrophication, primarily driven by increasing pollutant inputs, resulting in persistent algae blooms. Algae blooms in the lake have adverse effects on drinking water quality, habitats, and the ecosystem. In response to this challenge, both researchers and the Chinese Government have proposed various approaches to restore the lake. The government has taken steps such as shutting down polluting industries, prohibiting phosphorus-bearing detergents, and constructing water treatment plants. Researchers have developed methods like sediment dredging to combat eutrophication. Despite these efforts, the Lake Taihu water quality has not significantly improved. To address this, an engineering solution involved transferring water from the Changjiang River to Taihu Lake to dilute and flush out contaminations.

Water enters the lough via the Wangyuhe River and exits through the Taipuhe River (Fig. 1). Previous studies on the water transfer effects in lakes have produced conflicting results. While Wu (2008) demonstrated that water transfer enhances the water ecology and quality of Taihu Lough, Hu et al. (2010) argued that water divert could improve nitrogen and phosphorus content if the influent river has higher concentrations than certain areas within the lake. Effectively assessing the effect of water divert involves monitoring the water age of the lake using EFDC software.

Despite the advantages of the water diversion project, there are still some pressing problems:

1. What are the changes in water age throughout the entire lough or particular (polluted) areas within it?
2. In order to balance cost savings and water quality improvement, what is the best amount of water entering the Changjiang River and the optimal amount of water coming out of the pumping station?
3. What are the ecological and environmental impacts on both the "receiving" and "donating" systems carrying water?

To address these defiances, the following objectives were assayed:

1. 3 dimensional numerical model EFDC (environmental Fluid Dynamic Code) was used to study the effects of wind and the Changjiang River inflow on the environmental benefit of Taihu Lough.
2. Through multi-objective optimization, the optimal diversion flow of the original Changjiang River water was quantified on the premise of taking into account environmental benefits and economic costs.
3. Assess the effects of water diverted from Changjiang River Diversions on algal bloom dynamics.

The Impact of Water Diversion Works on the Whole Lough

The mean water age of the whole lough decreases piecemeal with the increase of the water diverted from the Changjiang River, and its variation range is closely related to the amount of incoming water. Under the southeasterly wind scenario, the water age decreased significantly from three hundred and one days to one hundred and seventy-nine days, and the inflow increased from fifty m^3/s to two hundred m^3/s. It is worth noting that when the input flow is low, the average water age change rate of the bay is higher. With the increase of transfer inflow from fifty to one hundred m^3/s, one hundred to one hundred and fifty m^3/s and 150 to two hundred m^3/s (scenario 1) in the southeast wind direction, the mean water age of the lough decreased by 26.14%, 16.20%, and 15.44%, respectively. This suggests that there was limited improvement in water exchange from one hundred and fifty m^3/s to two hundred m^3/s due to the initial Changjiang River diversion.

Wind direction is an important factor affecting the water age of lakes. The mean variation of water age was less than thirteen days in four wind directions: northwest, southwest, northeast, southeast, east, west, and no wind. However, under the south and north winds, the change is over fifty days. Furthermore, from the specific inflow point of view, the north wind scenario has the highest water age, followed by the south wind scenario, and the other wind direction has the lowest water age. For example, at an input flow of 200 m^3/s, the water ages of north, south and other winds are about 255, 227, and 173 days. This means that the influence of north and south winds on water transport performance is not significant.

Considering the combined effect of water divert and winds, water age fluctuated less with lower flow rates of diverted water compared to higher flow rates under various wind directions. Precisely, the averaged water age remained at 300 days with a fifty m^3/s inflow in all wind directions. But in higher flow rates of two hundred m^3/s, water age ranged from one hundred and fifty to two hundred and eighty days. Hence, wind directions had a substantial effect on the averaged water age of large discharge in the Yangtze River.

Effect of Water Divert Projects in the Lake

Water age is a crucial timescale for understanding the biogeochemical and hydrodynamic processes in lakes like Taihu Lough (Li et al. 2011). Younger water ages favor water exchange, nutrient dilution, and algae removal, while older ages can promote algae blooms. In this study, we found that Lake Taihu's water age was influenced by transferred water amounts and wind patterns, primarily due to its size and shallowness.

Throughout the lake, the southeasterly winds that prevail in summer facilitate the exchange of water most efficiently. This highlights the importance of considering wind patterns during the implementation of water diversion projects. In specific lake areas, the influx rate and the effect of wind are different because of the water transport route, coastline, topography, and other regional characteristics. The extent of the impact relies on the interaction of the transferred wind and water within each lough district.

In the main general, under different wind situations, the diversion of water from the former Yangtze River mainly improved the water interchange in some lakes, for example, Gonghu Bay, central and eastern outer margin areas. This indicates that the area near the Yangtze River channel is more effected by the inflow of the Changjiang River than the wind. However, in the most polluted areas, the impact of wind direction on water transfer path is greater than that of the water divert. The lake is a large, shallow lough, and this heterogeneous effect within it differs from smaller loughs such as Moses Lake and Green Lake in Washington, which show a more uniform and efficient interchange rate throughout the lough (Hilt et al. 2011).

Enhanced Changjiang River water diversion projects, including Xingou and Meiliang pumping stations, have improved water exchange and mitigated environmental and ecological risks in Meiliang Bay. Thanks to those pump stations, Meiliang Bay witnessed reduced algal blooms directly correlated with younger water ages.

Impact on Aquatic Ecosystem

Inter-basin water transfer projects pose significant ecological challenges, including alterations in water quality, eutrophication, algae blooms, and species invasions. These projects offer positive effects, such as shorter water ages, generally elevated water levels reducing internal load release and sediment heavy suspension, and dilution of nutrient concentrations in the water column. However, negative consequences are evident, including increased spatial heterogeneity in water age, potentially worsening algal growth and outbreaks, and the transfer of pollutants from one area to another without effective removal from the machinery, thereby degrading the water property.

One noteworthy example of inter-basin water transfer is the Yangtze River water transfer projects, which have garnered significant attention for their interaction between the "receiving" (the Taihu Lake) and "donating" (the Changjiang River) ecosystems. The diminish in Chl-a denseness and water quality betterment, to some extent, following water transfer is typically a short-range response unless nutrient concentrations in the diverted water are reasonably reduced. Sewage flowing out of Lake Taihu through the River Taipu and the new pumping station inclusions high concentrations of algae and nutrients that can effect besieging aquatic ecosystems, embodying the Changjiang River.

Similar projects worldwide have been subject to investigations regarding their impact on besieging aquatic ecosystems. For instance, the Snowy hill Scheme in Southeast Australia, before the environmental affect assessment, has raised concerns about potential ecological impacts on both recipient and donor systems, including salt wedging and flow reductions in the contributor, leading to changes in sediment carry in basins and aquatic machineries (Davies et al. 1992). The diversion project of South to North Water in China has also generated environmental issues, disrupting the vulnerable ecological balance in China North. This includes sedimentation in river mouths, seasonal drying of river courses, severe land subsidence, and the development of groundwater depression cones (Liu and Zheng 2002).

Hence, when planning the project of transferring water on a large scale between basins, it is essential to conduct comprehensive studies that consider the technical aspects, and the potential ecological affects, while factoring in social and economic effects.

3.8 *Policy*

China's Water Policy Experience in Dealing with Water Contamination

The importance of water as a fundamental component of life is widely recognized (European Environment Agency 2018). However, this importance holds true when water quality is not compromised and meets specific standards. Regrettably, water bodies across the globe are contending with severe pollution from various sources, resulting in a significant number of deaths due to waterborne diseases. Many countries have experienced varying degrees of water pollution. In Europe, approximately sixty percent of water bodies are made impure, while in the USA, this percentage is roughly 40% (European Environment Agency 2018). The congeneric degree of water contamination is often related to elements such as standard of advancement, population growth, and national or international water pollution control lawmaking. The concluding is of paramount importance as it shapes people's understanding of how to live in ways that safeguard their environment. The implementation of water policies has played a crucial role to encourage countries to dramatically improve the quality.

Among countries, India established "The Andhra Pradesh Pollution Control Board" to manage environmental contamination, incorporating various waste treatment facilities. Additionally, they achieved the closure of approximately one hundred and sixteen polluting industries. The first national standard limiting industrial water pollution has been issued by the Environmental Protection, which has inhibited around one billion pounds of contamination flow into water in the USA (Environmental Protection Agency 2018). In China, eutrophication stands as one of the principal concerns regarding water contamination. Like other countries, China's water policy for pollution control has implemented numerous measures and actions that have yielded significant influence.

3.8.1 Long River System

The River Chief system in China constitutes a series of standards aimed at improving water quality and preserving water resource completeness (Smith 2018). When Lake Tai suffered a severe algae bloom, the foundation of the system traces back several years, significantly deteriorating the lake's water quality and endangering the drinking water supply for approximately five million residents in Wuxi city (Smith 2018). To address this water quality issue, the local government assigned officials to monitor water quality measures in specific areas. The successful outcomes of the initiative prompted nearby cities to adopt a similar approach in the subsequent years (Smith 2018). In 2016, China's central government launched a nationwide "River Long" program to control and manage solid waste discharge (Ministry of Water Resources of China 2018).

River Chief System

In response to the escalating water pollution in China, local authorities have implemented various measures aimed at monitoring rivers and lakes. Inspired by the successful initiatives undertaken by these local authorities to manage and safeguard their water-bodies, the river head system has garnered significant heed in China. Recognizing the critical importance of water safety, the CPC Central Committee and the State Council have prioritized this issue. To that end, the Central Committee of the Communist Party of China issued the Opinions on Comprehensively Promoting the River Chief System, which outlines a global commitment to green development through advanced water treatment and assigns local authorities the responsibility for managing rivers and lakes (Ministry of Water Resources of China 2017b).

3.8.2 Action Plan for Water Pollution Control

Law of the People's Republic of China on Restraint and Control of Water Contamination

China has established a comprehensive legal framework to address water pollution through a series of seven chapters as outlined by the Ministry of Ecology and Environment of the People's Republic of China (1984). These chapters encompass:

Chapter 1: Comprising five articles, this section defines general provisions;

Chapter 2: The Measures for the Management of Water Ambient Quality have three provisions, which stipulate the overall standards for water environmental quality and the emission of water contamination;

Chapter 3: Encompassing seventeen articles, this chapter outlines the superintendence and management of water resource prevention and control;

Chapter 4: Comprising 14 articles, this section addresses the avoidance of surface water contamination;

Chapter 5: Including five articles, it focuses on preventing groundwater contamination;

Chapter 6: Devoted to legal obligation, it contains thirteen articles;

Chapter 7: This final chapter consists of 3 articles, providing supplemental provisions.

Water Pollution Control Action Plan of China

In response to its economic maturation, China has formulated an array of measures to address the challenge. The State Council of China has set ambitious goals, envisioning substantial improvements in water quality by 2020 and a reduction in the advancement of groundwater pollution nationwide (The State Council of China 2015). Looking ahead, the government aims to implement long-term ecological management strategies to achieve the concept of "the virtuous circle" within the ecosystem.

3.8.3 Implementing Most Strictest Water Resources Administration System—3 Redlines

China implemented a rigorous assessment of its water resources management system. This assessment relied on specific indicators that measured the system's ability to achieve its goals and objectives in 2016 (Ministry of Water Resources of China 2017a). The assessment covered approximately thirty-one provinces, municipalities, and autonomous regions directly under the central government's jurisdiction, with the following key outcomes endorsed by the State Council (Ministry of Water Resources of China 2017a): 1. All thirty-one provinces met the judgment standards, with 5 of them achieving a high-grade rating; 2. Over the period from 2015 to 2016, there was a 7.2% reduction in water use intensity per GDP and a 7.6% decrease in industrial value; 3. Guidelines outlining reforms for water usage in households and industries at various administrative levels (county, province, and borough) were issued.

3.8.4 Protection of the Changjiang River

The Changjiang River region encompasses around nine economic provinces and two municipalities, spanning approximately one-fifth of China's land area and providing a home to approximately 600 million people (The State Council of The People's Republic of China 2016). In 2014, the Chinese government introduced strategies and measures aimed at advancing the advancement of this economic zone, with a strong emphasis on ambient protection and water resources management. These strategies are commonly referred to as "green development" and "golden waterway" initiatives (The State Council of China 2016).

Green Advancement

The scheme outlines a series of actions aimed at achieving specific goals related to water management and environmental protection. These actions include: (a) implementing effective management without regard to market factors or administrative boundaries. (b) Ban some polluting enterprises along the Yangtze River; (c) Setting emission standards for pollutants and establishing local compensation mechanisms to encourage environmental protection; (d) Strengthen water pollution control throughout the basin by treating wastewater from major tributary urban areas (Government of China 2016); (e) Promoting flood control systems to support ecological restoration projects (Government of China 2016); (f) Formulate effective measures to govern natural catastrophes in the Changjiang River region; (g) Establishment of a dedicated funding beginning for the secure wetlands (Government of China 2016).

Golden Water Way

In pursuit of economic prosperity and development, the government has set forth specific targets, including those related to urbanization and industrial upgrading, encapsulated in the initiative. These policies encompass: (a) The restoration of the watercourse; (b) The low-carbon development infrastructure (such as roads and railways) by 2020; (c) The regulation of all waterway construction (Yangtze River), along with the implementation of specific standards for ships; (d) Increased investment allocation for research and innovation; (e) The establishment of three economic hubs along the waterway to facilitate urbanization; (f) Over the long term, the implementation of advanced industries along the waterway to replace inefficient ones.

References

Bloomberg News (2017) World's largest water diversion plan won't Quench China's Thirst—Bloomberg. In: Bloom. News

Bo S, Liu C, Yun Z et al. (2013) Urban waste disposal and its impact on groundwater pollution in China. J Groundwater Sci Eng 477–504

Cauwenberghe LV, Devriese L, Galgani F et al (2015) Microplastics in sediments: a review of techniques, occurrence and effects. Mar Environ Res 111:5–17

Chen J, Lin D, Deng B (1989) A study on heavy metal partitioning in sediments from Poyang Lake in China. Hydrobiologia 176–177:159–170

Chen ZH, Chen NW, Wu YQ et al (2014) Sediment-water flux and processes of nutrients and gaseous nitrogen release in a China River Reservoir. Huan Jing Ke Xue 35:3325–3335

Chinese Government (2016) Outline of the thirteenth five-year plan for national economic and social development of the People's Republic of China

Coveney MF et al (2003) Nutrient removal from eutrophic lake water by wetland filtration. Ecol Eng 19(2):141–159

Currell MJ, Cartwright I, Bradley DC et al (2010) Recharge history and controls on groundwater quality in the Yuncheng Basin, north China. J Hydrol 385:216–229

Davies BR, Thoms M, Meador M (1992) An assessment of the ecological impacts of inter-basin water transfers, and their threats to river basin integrity and conservation. Aquat Conserv 2:325–349

Deqi L (2005) Identification to effects of eco-restoration for water bodies of eutrophication. East China Normal University

Ding-Xin WU et al (2010) Control of water blooms of urban eutrophic lake with compound microorganisms in restoration project. Environ Sci Technol 33(7):150–154

Environmental Protection Agency (2018) What has the U.S. EPA (Environmental Protection Agency) accomplished in its history?

European Environment Agency (2018) Water use and environmental pressures

Fabbricino M, Petta L (2007) Drinking water denitrification in membrane bioreactor/membrane contactor systems. Desalination 210(1):163–174

Fan CH, Chang M, Zhang YC (2016) Spectral characteristics of dissolved organic matter (DOM) derived from water and sediment in normal flow period of the intersection zone of Jing River and Wei River. Spectrosc Spectral Anal

Fang D, Zhang R, Zhou L et al (2011) A combination of bioleaching and bioprecipitation for deep removal of contaminating metals from dredged sediment. J Hazard Mater 192:226–233

Fischer WJ, Dietrich DR (2000) Pathological and biochemical characterization of microcystin-induced hepatopancreas and kidney damage in carp (*Cyprinus carpio*). Toxicol Appl Pharmacol 164(1):73–81

Frische K, Schwarzbauer J, Ricking M (2010) Structural diversity of organochlorine compounds in groundwater affected by an industrial point source. Chemosphere 81:500–508

Gómez MA, González-López J, Hontoria-García E (2000) Influence of carbon source on nitrate removal of contaminated groundwater in a denitrifying submerged filter. J Hazard Mater 80(1):69–80

Guo H, Zhu J, Yang Y (2008) Research status and development of technologies for controlling agricultural non-point source pollution. Trans Chin Soc Agric Eng

GuoXiang W, Peimin P (1999) Influence of some artificial controls on eutrophic algal population dynamics. Environ Sci 2:71–74

Hang C et al. (2010) Application of bio-film technology in polluted surface water bio-remediation. J Zhejiang Wat Con Hydr Coll 22(1):43–46

Hilt S, Köhler J, Kozerski HP, Nes EH, Scheffer M (2011) Abrupt regime shifts in space and time along rivers and connected lake systems. Oikos 120(5):766–775

Hongmei M, Tian Z, Peng Y (2017) Advances in in situ ecological restoration of eutrophic water bodies. J Chengdu Textile Coll 34(4):156–159

Hu WP, Zhai SJ, Zhu ZC, Han HJ (2010) Impacts of the Yangtze River water transfer on the restoration of Lake Taihu. Ecol Eng 34:30–49

Hua ZL, Gu L, Xue H, Liu XD (2008) Assessing indicators for water diversion based on improving water quality of shallow lakes. J Lake Sci 5:623–629 (in Chinese)

Huang T et al. (2006) The onsite experiment study of enhanced bio-contact oxidation process for pretreatment of Luanhe River water. Acta Scientiae Circumstantiae

Huang T, Jia S, Cong H (2008) Experimental study on the water-lifting aerator enhanced chemical oxidation process for polluted raw water. Water Wastewater Eng

Huang T (2016) Water pollution and water quality control of selected chinese reservoir basins. Handbook Environ Chem

Huiguang G, Pijing M (2000) Treatment framework and estimated investment of Dianchi Lake. Environ Sci Surv 19(a08):109–112

Ito C (2005) Urbanization and water pollution in China. Australian National University

Ji F, Wang G, Han R et al (2016) Vertical distribution and pollution characteristics of nutrient in sediment of rural malodorous black river in Wuxi City, China. Chin J Environ Eng 10:3462–3470

Jin T, Chao L, Li W et al. (2009) Advances in heavy metal pollution in marine environment. Fisheries Sci

Le Rouzic B, Thiébaut G, Brient L (2016) Selective growth inhibition of cyanobacteria species (*Planktothrix agardhii*) by a riparian tree leaf extract. Ecol Eng 97:74–78

Li YP, Acharya K, Yu ZB (2011) Modeling impacts of Yangtze River water transfer on water ages in Lake Taihu, China. Ecol Eng 37:325–334

Li D, Zuo et al. (2013) Disposal of chemical contaminants into groundwater: viewing hidden;environmental pollution in China. Environ Earth Sci 70:1933–1935

Linfeng L et al (2008) Potential of constructed wetlands in treating the eutrophic water: evidence from Taihu Lake of China. Bioresour Technol 99(6):1656–1663

Liu C, Zheng H (2002) South-to north water transfer schemes for China. Int J Water Resour D 18(3):453–471

Mei LX et al. (2000) Control of algae bloom in eutrophic water by effective microorganisms. Actaentiarum Naturalium Universitatis Sunyatseni 39
Meizi L, Liyi Y, Shaoqin Z (1992) Population variation of phytoplankton in West Lake of Hangzhou before and after diluting sewages with empting into river water and effect of controlling eutrophication. Chin J Appl Ecol 3(3):266–272
Minh NH, Minh TB, Kajiwara N et al (2007) Pollution sources and occurrences of selected persistent organic pollutants (POPs) in sediments of the Mekong River delta, South Vietnam. Chemosphere 67:1794–1801
Ministry of Ecology and Environment of the People's Republic of China (1984) Law of the People's Republic of China on prevention and control of water pollution
Ministry of Ecology and Environment of the People's Republic of China (2017) Report on the State of the Ecology and Environment in China. http://english.mee.gov.cn/Resources/Reports/soe/SOEE2017/201808/P020180801597738742758.pdf
Ministry of Water Resources of The People's Republic of China (2018) China rolls out "river chief" scheme nationwide
Ministry of Water Resources of The People's Republic of China (2017a) Water policies, laws and regulations in China
Ministry of Water Resources of the People's Republic of China (2015) Water Resources Management and Protection in China. http://www.mwr.gov.cn/english/mainsubjects/201604/P020160406507020464665.pdf
Ministry of Water Resources of the People's Republic of China (2017b) China water resources bulletin
Ministry of Water Resources of The People's Republic of China (2017a) Press conference was held by Ministry of Water Resources on the assessment implementation of the most stringent water resources management system
Naumann MS et al (2015) Monitoring of coastal coral reefs near Dahab (Gulf of Aqaba, Red Sea) indicates local eutrophication as potential cause for change in benthic communities. Environ Monit Assess 187(2):44
Nazzal Y, Ahmed I, Al-Arifi NSN et al (2014) A pragmatic approach to study the groundwater quality suitability for domestic and agricultural usage, Saq aquifer, northwest of Saudi Arabia. Environ Monit Assess 186:4655–4667
Ni Z, Wang S (2015) Historical accumulation and environmental risk of nitrogen and phosphorus in sediments of Erhai Lake, Southwest China. Ecol Eng 79:42–53
Ning W, Gang Z, Yuan W (2004) Environmental protection dredging and its effects for intrinsic pollution of lake: case study of Nanhu Lake in Changchun. Res Environ Sci 17(2):34–37
Pigram JJ (2000) Options for rehabilitation of Australia's Snowy River: an economic perspective. Regul River 16(4):363–373
Qin (2006) Estimation of internal nutrient release in large shallow Lake Taihu, China. Sci China 49:38–50
Qiqi S, ChunPing Z, XingXiu YU, JianHua LI, YongKun Z, Yan G (2013) Best management practices of agricultural non-point source pollution in china: a review. Chin J Ecol 32:772–778
Rauret G, Galceran MT, Rubio R et al (2012) Factor analysis for assigning sources of groundwater pollution. Int J Environ An Ch 38:389–397
Rice EL (1984) Allelopathy, 2nd ed. Physiol Ecol
Rivas Z et al (2000) Nitrogen and phosphorus levels in sediments from tropical Catatumbo River (Venezuela). Water Air Soil Pollut 117(1–4):27–37
Rouabhia A, Baali F, Fehdi C (2010) Impact of agricultural activity and lithology on groundwater quality in the Merdja area, Tebessa, Algeria. Arab J Geosci 3:307–318
Runan T et al (2011) Nitrogen and phosphorus removal effects of different hydrophyte combinations under simulated eutrophic conditions. J Beijing For Univ 33(6):191–195
Sheffield VC et al (1989) Attachment of a 40-base-pair G + C-rich sequence (GC-clamp) to genomic DNA fragments by the polymerase chain reaction results in improved detection of single-base changes. Proc Natl Acad Sci U S A 86(1):232–236

Shumei T (2016) Comparison of several ecological restoration techniques for eutrophic water bodies. China Water Transp 16(7):113–116
Singer A et al (2008) A novel approach to denitrification processes in a zero-discharge recirculating system for small-scale urban aquaculture. Aquacult Eng 39(2):72–77
Singh D, Tiwari A, Gupta R (2012) Phytoremediation of lead from wastewater using aquatic plants. Int J Biomed Res 2(7):411–421
Smith S (2018) China's River chiefs—an "Ingenious" approach to water governance in China?
Sun JC, Liu JT, Ji xiang QI et al. (2015) Modern investigation, sampling and analysis technological system in the investigation of groundwater pollution in China. Acta Geoscientica Sinica
The Economist (2018) China has built the world's largest water-diversion project—a massive diversion. Econ
The State Council of The People's Republic of China (2016) Environmental protection top priority of Yangtze River economic belt development
The State Council the People's Republic of China (2015) China announces action plan to tackle water pollution
Tong C et al (2016) Different combination of fishes, benthic animals and aquatic plants control of eutrophication water bodies by mesocosm experiment. Chin J Environ Eng 10(10):5511–5520
Trauth R, Xanthopoulos C (1997) Non-point pollution of groundwater in urban areas. Water Res 31:2711–2718
Van de Bovenkamp J (2016) Developments and opportunities in water and environmental technology in China
Wang Y, Yang J, Liang J, Qiang Y, Fang S, Gao M et al (2018) Analysis of the environmental behavior of farmers for non-point source pollution control and management in a water source protection area in china. Sci Total Environ 633:1126–1135
Wen X, Du et al. (1998) Surface complexation model for the heavy metal adsorption on natural sediment. Environ Sci Technol 32:870–875
Wu H (2008) Study on key technologies of Yangtze-Taihu Water Diversion. China Water Resour 1:6–8 (in Chinese)
Xia J et al (2017a) Analysis of concepts, conditions and critical problems in environmental dredging. Res Environ Sci 30(10):1497–1504
Xia J, Zhang YY, Xiong LH, He S, Wang LF, Yu ZB (2017b) Opportunities and challenges of the sponge city construction related to urban water issues in China. Sci China Earth Sci 60:652–658
Xiang W et al. (2013) Effects of allelochemical extracted from water lettuce (*Pistia stratiotes* Linn.) on the growth, microcystin production and release of *Microcystis aeruginosa*. Environ Sci Pollut Res 20(11):8192–8201
Xiaopeng H (2017) Types of ecological revetments in river regulation. Port Waterway Eng 11:118–122
Xie P, Liu J (2014) Practical success of biomanipulation using filter-feeding fish to control cyanobacteria blooms: a synthesis of decades of research and application in a subtropical hypereutrophic lake. Sci World J 1:337
Xingyu JG, Kefu Z (2001) Mechanism of heavy metal injury and resistance of plants. Chin J Appl Environ B Iol 7(1):92–99
Xu M, Yao RH, Song LL (2015) Primary exploration of general plan of the urban black-odor river treatment in China. Chin J Environ Manage 74–78
Xubao L (2013) Application of bioremediation technology in urban water environment management. Urban Roads Bridges Flood Control 9:177–178
Yang JF, Ying GG, Zhao JL et al (2010b) Simultaneous determination of four classes of antibiotics in sediments of the Pearl Rivers using RRLC-MS/MS. Sci Total Environ 408:3424–3432
Yang Z et al. (2010a) Remediation of eutrophic water by ecological floating bed shanghai chemical industry. 35(11):1–4
Yao HY (1998) A history of the grand canal. China Water Conservancy and Hydropower Press (in Chinese)

Yong C, Congjun S (2009) Application of ecological floating beds to water restoration and its design. Environ Sci Technol 32(2):121–124

Zhang J, He M, Lin C et al (2010) Phenanthrene sorption to humic acids, humin, and black carbon in sediments from typical water systems in China. Environ Monit Assess 166:445–459

Zhaohai B, Jie L, Hao Z, Velthof GL, Oene O, Dave C et al (2018) Designing vulnerable zones of nitrogen and phosphorus transfers to control water pollution in china. Environ Sci Technol 2018(52):8987–8988

Zhen Z, Xiaosong S, Boqing G (2011a) Discussion on ecological repair technology of water body eutrophication. Heilongjiang Environ J 35(1):69–71

Zhong J (2009) Environmental effect of sediment dredging in lake (I): the role of sediment dredging in reducing internal phosphorous release. J Lake Sci 21(1):84–93

Zhou X (2009) Water pollution control in China: review of laws, regulations and policies and their implementation

Zilu H (2011) Management and restoration project effect evaluation of lakeside wetland ecosystem, Nanjing Forestry University

Zou W, Yuan L, Zhang L (2013) Analyzing the spectral response of submerged aquatic vegetation in a eutrophic lake, Shanghai, China. Ecol Eng 57:65–71

Sustainable Management of Water Resources in a Semi-arid River Basin Under Climate Change: A Case Study in South Africa

Yali E. Woyessa

1 Introduction

The shortage of water availability for drinking, sanitation, and agricultural practices coupled with extreme events such as floods and droughts severely impact most of the world's population in general and sub-Saharan Africa in particular. The National Water Resources Strategy of South Africa puts emphasis on the fact that water resource is central to development and poverty alleviation (DWA 2013). However, decision-makers face many challenges in their effort of ensuring the sustainable and equitable use of the available water resources. Some of these challenges include rapidly growing population, changing economies, and climate change.

The problem of water resources availability is likely to deepen in the future. It is projected that, by 2050, the Southern African region will be generally drier (DEA 2017). In general, for South Africa, based on the evidence of the past five decades there is a tendency toward a rise in temperature (MacKellar et al. 2014). Maximum and minimum temperatures have also shown significant increases annually, and in almost all seasons. High temperature extremes have increased significantly in frequency, and low temperatures have decreased significantly in frequency annually and in most seasons across the country (DEA 2017). Recent climate projections suggest a drop of up to 10% in precipitation in most of Southern Africa by 2050 (Levina 2006). On one hand, it is estimated that by the year 2025 almost one-half of the world population will be living in water-stressed regions (World Water Council 2000). On the other hand, rapid population increase, industrialization, and pollution are subjecting further strain on the available and diminishing fresh water resources. Within this context, there is a general understanding that development has to be sustainable, adaptable, and resilient to these global changes such as climate change.

Y. E. Woyessa (✉)
Department of Civil Engineering, Central University of Technology, Free State, Bloemfontein 9300, South Africa
e-mail: ywoyessa@cut.ac.za

Y. Li et al. (eds.), *BRICS Countries: Sustainable Water Resource Management and Pollution Control*, https://doi.org/10.1007/978-981-99-9581-3_8

Resilient development involves initiatives that will deliver benefits under all potential future circumstances related to environmental changes (AMCOW 2012). Therefore, this calls for innovative strategic approaches to improve water resources availability and resilient development.

In South Africa, traditionally, water supplies have increased through the developments of new reservoirs and other water development schemes. However, the current status of river basin development shows that all basins are committed and there are no more options for further development (Turpie et al. 2008). This implies that there is a need to explore other options for increasing and conserving water supplies. One of these options could be effective rainwater management, but this requires understanding the hydrological dynamics under different climate change scenarios.

The hydrological balance of any river basin is directly and indirectly influenced by the spatial and temporal distribution of land-use and land-cover changes as well as climate condition. Changes in land use can modify crucial hydrological processes, such as evapotranspiration (ET) and groundwater recharge. Upstream land-use change could bring about a significant on-site effect on water security at watershed level and off-site effect on downstream users (Woyessa et al. 2006, 2011). Moreover, the effect of climate change could further exacerbate the water security problem through extreme weather conditions, such as drought and flooding. It is therefore important to develop scenarios on possible climate and land use change and investigate the possible impacts of these changes on the hydrological processes and water resources availability. The main aim of this paper is to investigate the impact of climate change scenarios on water resources to explore management options and adaptation strategies.

2 Methodology

2.1 Description of the Study Area

The research was conducted in a semi-arid river basin in the central region of South Africa (Free State Province), known as the Modder River basin (Fig. 1). The Modder River basin has a total area of 1.73 million hectares (17,300 km^2). It is divided into three sub-basins, namely the Upper Modder, the Middle Modder, and the Lower Modder.

Socio-economic Factors

The Free State province is one of the nine provinces and is centrally located within South Africa. The last census conducted in 2011 shows that the province has a population of 2.7 million people (Table 1). According to the provincial government, the province is sparsely populated at 22 people per km^2 and approximately 29% of individuals in the Free State province live in rural farming areas of the province (Free State Provincial Government n.d).

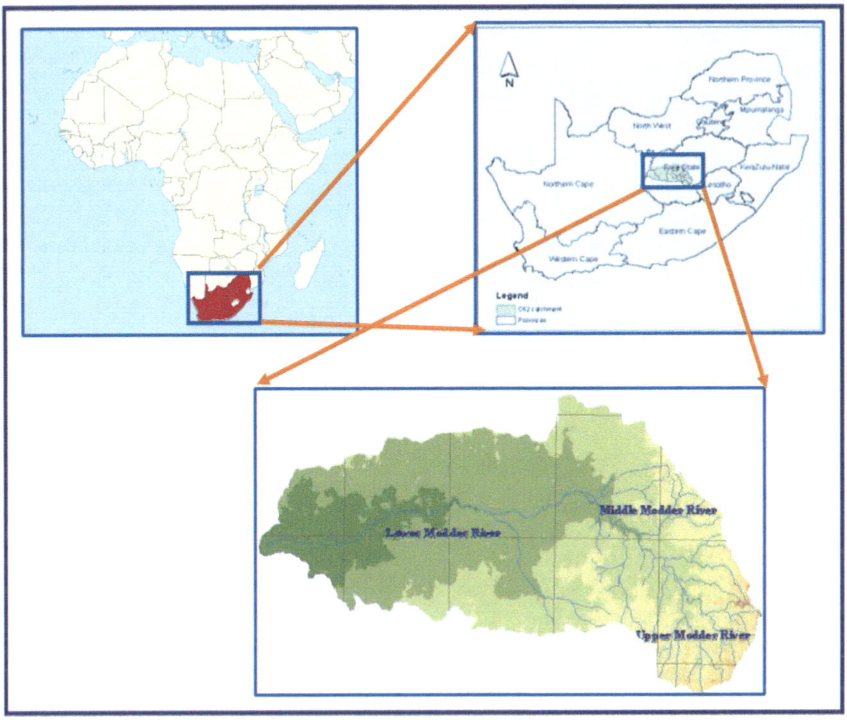

Fig. 1 The study site

The Free State is losing ground in terms of population development in South Africa. The Free State population was 2.633 million in 1996 and increased by 73 271 to 2.706 million in 2001. The provincial population further increased by 38 815 to 2.745 million according to the 2011 census. This data alone shows that the population of the Free State showed a decreasing rate between 1996, 2001, and 2011 (Puukka et al. 2012).

Specific information on the Modder River basin's socioeconomic and demography is very limited. However, some of the information could be inferred from reports compiled on the Free State Province as well as the Motheo District Municipality where much of the Modder River basin is located.

The Modder River basin comprises a major city and two townships within the Free State Province, namely the City of Bloemfontein and the townships of Botshabelo and Thaba Nchu. These three cities account for the biggest proportion of the province's population with a combined figure of 747,431 according to the national population census of 2011. However, it was reported that the population has declined by 22,382 in the years that followed the official census (MLM IDP 2015/16). This is also in line with the OECD report, which found a decreasing trend of population growth in the Free State province.

Table 1 Population of the Free State and other provinces of South Africa

Province	Census 1996	Census 2001	Census 2011	Population estimate 2014	% Change 1996–2014	% Change 2001–2014	% Change 2011–2014	2014 Share of population
Eastern Cape	6,147,244	6,278,651	6,562,053	6,786,900	10.4%	8.1%	3.4%	12.6
Free State	2,633,504	2.706,775	2,745.590	2,786.800	5.8%	3.0%	1.5%	5.2
Gauteng	7,834,125	9.388,854	12,272,263	12,914,800	64.9%	37.6%	5.2%	23.9
KwaZulu-Natal	8,572,302	9,584,129	10,267.300	10,694,400	24.8%	11.6%	4.2%	19.8
Limpopo	4,576,566	4.995,462	5.404,868	5,630.500	23.0%	12.7%	4.2%	10.4
Mpumalanga	3,123,869	3,365,554	4,039,939	4,229,300	35.4%	25.7%	4.7%	7.8
Northern Cape	1,011,864	991,919	1,145,861	1,166,700	15.3%	17.6%	1.8%	2.2
NorthWest	2,727,223	2,984,098	3,509,953	3,676.300	34.8%	23.2%	4.7%	6.8
Western Cape	3,956,875	4,524,335	5.822,734	6,116,300	54.6%	35.2%	5.0%	11.3
South Africa	40,583,572	44,819,777	51,770,561	54,002.000	33.1%	20.5%	4.3%	100

In terms of the economic situation in the Modder River basin, the average household income per annum in Bloemfontein, Thaba Nchu, and Botshabelo are R76,611; R32,105, and R24,828, respectively. Approximately 15% of the household earn between R6,000 and R18,000 (Woyessa et al. 2011).

Land ownership in the basin is both private and communal. According to Woyessa et al. (2006) 35% of the upper Modder River basin was occupied by communal farmers. The area occupied by the communal farmers is marginal for crop production as well as largely unproductive due to insufficient rainfall for rain-fed agriculture. Most smallholder farmers in the communal areas report that they lack farming knowledge, especially regarding crop production (Kundhlande et al. 2004). A significant portion of land in the Upper Modder River basin is communally owned, while commercial farmland dominates in the middle and lower part of the Modder River basin. In the communally owned land part, land is distributed and allocated by a Tribal Chief (Kundhlande et al. 2004).

Biophysical Factors

Biophysical factors, such as climate, land use, soil, and topography are some of the major factors that greatly affect runoff that will be generated from a specific watershed (Peugeot et al. 1997; Jain et al. 2004). Rainfall and evapotranspiration (ET) are the two main components of the climate that have major impact on the water balance of a watershed. High rainfall contributes to increased runoff while high ET reduces the amount of overland flow (surface runoff), which is considered as blue water (Jewitt et al. 2004). Similarly, soil properties, such as texture and infiltration rate affect the amount of surface runoff considerably. For instance, soils with high silt and clay content and soils with low infiltration rate contribute to high surface runoff, whereas soils with high sand content and soils with high infiltration rate reduce runoff and stream flows.

Climate

The mean annual rainfall of the study area is 551 mm. The mean monthly rainfall distribution is also shown in Fig. 2. As can be observed from the Figure, the rainy season consists of January to April and September to December, leaving the months of May to August as dry winter season. The highest maximum temperature varies between 28°C and 31°C while the lowest minimum temperatures vary between -1°C and 1°C.

Soil

According to the soil map and GIS database obtained from the Institute of Soil Climate and Water (ISCW), seven soil series are identified in the Modder River basin (Fig. 3). The texture of this soil is 100% sandy clay loam to sandy clay (SaClLm to SaCl). The Middle Modder River basin is predominantly covered by soil series whose texture is dominated by sand to loamy sand (Sa to LmSa). The Lower Modder River basin is largely represented by soil series with a dominant texture of sandy clay loam (SaClLm).

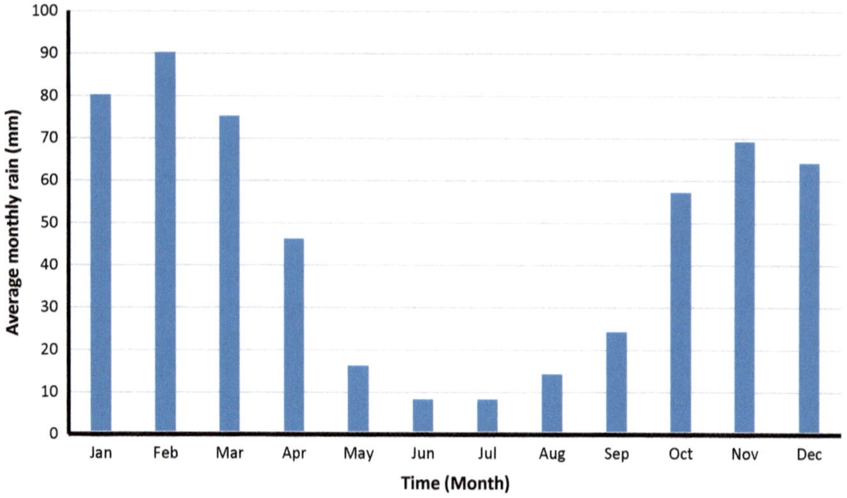

Fig. 2 Long-term average monthly rainfall

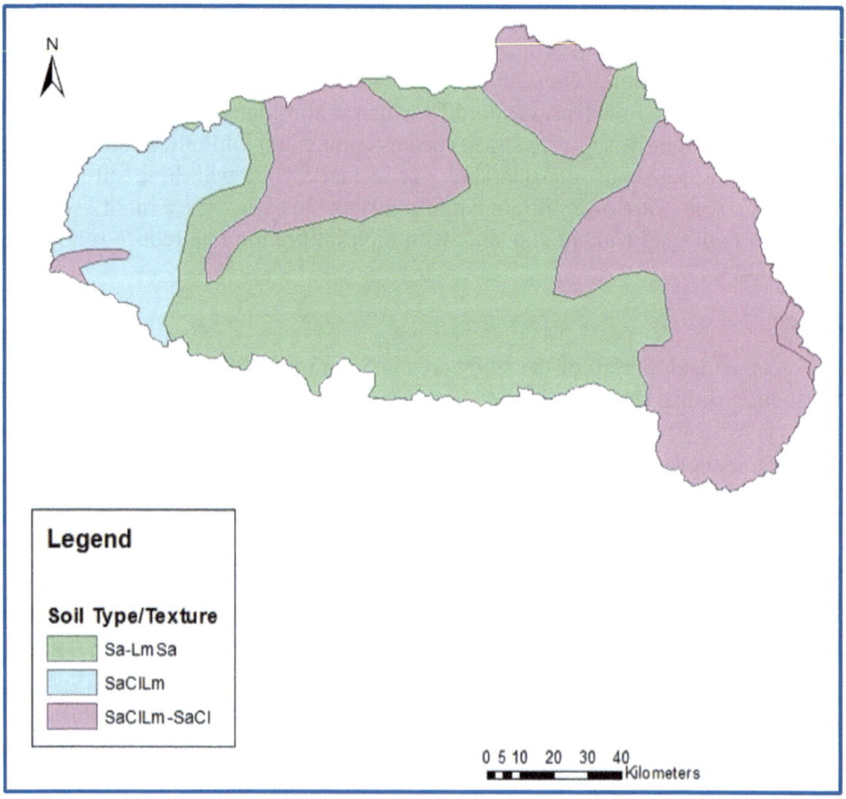

Fig. 3 Soil types and its distribution in the Modder river basin

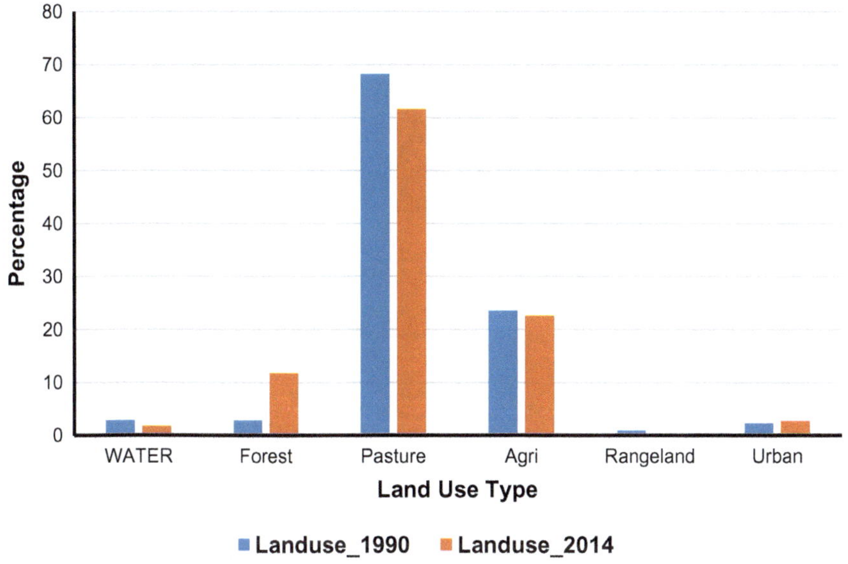

Fig. 4 Percentages of land use types for 1990 and 2014

Land Use

Land use/land cover data was obtained from the Department of Environmental Affairs of the Republic of South Africa (DEA 2017) for the years 1990 and 2014. The summary of the land use types, as per the class names recognized by SWAT, is given in Fig. 4. The figure shows percentage of land use types and possible changes during this 14 years' period. As can be observed from the figure, the two major land use types are pasture and agriculture showing small changes (decreasing trend) during these two periods. There is also a significant increase in forest cover and a slight increase in urbanization.

Topography

The topographic data or digital elevation model (DEM) for the study area was obtained from the USGS Earth Explorer, the Shuttle Radar Topography Mission (SRTM) 1Arc-Second Global (https://earthexplorer.usgs.gov) with a spatial resolution of 30 m. Figure 5 shows the DEM as well as the stream network of the river basin. The altitude ranges from the lowest of 1065 m.a.s.l. to the highest of 2129 m.a.s.l.

2.2 Climate Change Modeling

Recognizing the availability of limited studies on regional climate downscaling (RCD), the World Climate Research Program formed a task force, which was

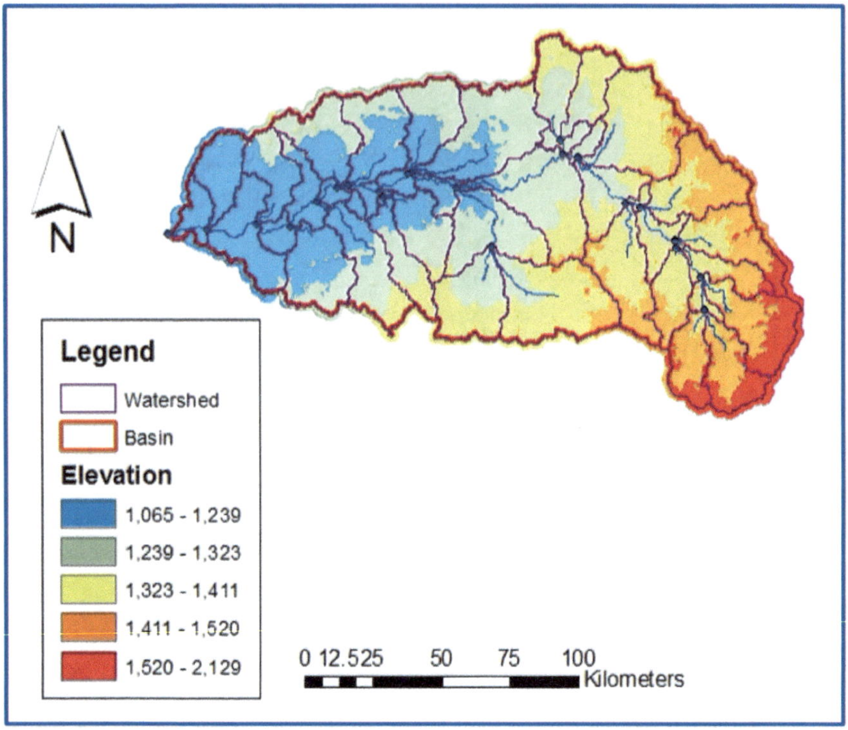

Fig. 5 DEM of the modder river Basin

mandated, among others, to develop a framework to evaluate and possibly improve RCD techniques for use in downscaling global climate projections. As a result of the first activities of this task force, and in consultation with the broader scientific community, a framework was initiated which is known as the Coordinated Regional Climate Downscaling Experiment (CORDEX) (Fig. 6). The downscaled climate data from CORDEX covers the duration from 1951 to 2100, of which 1950–2005 for retrospective run and 2020 to 2100 for prospective run, and two greenhouse emission scenarios, namely RCP4.5 and RCP8.5. The spatial resolution for Africa is reported to be 50 km grid spacing.

The climate change scenarios (precipitation, and minimum and maximum temperatures) were generated using eight GCM driven models for CORDEX dataset. The temperature change scenario was captured through potential evapotranspiration as outputs of the SWAT hydrological model. Historical simulation of precipitation was conducted in order to establish how well the models were able to predict historical values.

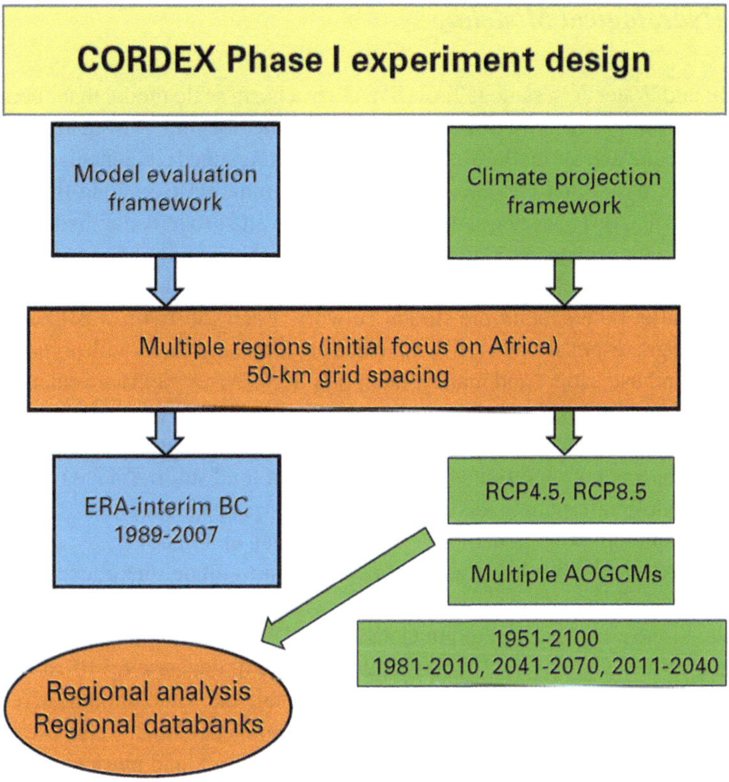

Fig. 6 Schematic depiction of the first phase CORDEX experiment set-up (Giorgi et al. 2009)

Type of Data and GCM Models

As described above, the use of dynamical downscaled data (CORDEX) was considered and a sub-set of eight GCMs was selected. Furthermore, daily data of precipitation, and minimum and maximum temperatures for Retrospective Run of 55 years (1951–2005) and Prospective Run of 80 years (2020–2100) were analyzed. The two greenhouse gas emission scenarios, RCP4.5 and RCP8.5, were considered, which are said to be the highest priority global model simulations within CIMP5. The eight climate models used for the datasets were: CNRM-CM5, IPSL-CM5A-LR, MIROC5, MPI-ESM-MR, CanESM2, CSIRO-Mk3-6-0, GFDL-EMS2M, and NorESM1-M.

2.3 *Hydrological Modeling*

The Soil and Water Assessment Tool (SWAT) is a basin scale model that operates on a continuous daily time step and is designed to predict the impact of management on water, sediment, and agricultural chemical yields in ungauged catchments. The model is physically based and capable of continuous simulation over long-time periods (Gassman et al. 2007). It integrates weather, surface and groundwater hydrology, soil properties, plant growth, and land management practices to model processes within a watershed (Arnold et al. 1998). In SWAT, watersheds are divided into sub-basins based on outlet points along the stream network. Each sub-basin was then divided into hydrologic response units (HRUs), which are defined as areas with homogeneous soil type, land use, slope, and management practices. Water yields are calculated for each HRUs and summed up to determine the total sub-basin outputs, which in turn contribute to the total watershed yield output. The HRUs are not spatially defined within sub-basins but represented as percentages of total sub-basin area. Therefore, SWAT includes both spatially distributed parameterization at the sub-basin scale and lumped parameterization at the HRU scale (Gassman et al. 2007).

The SWAT model has been successfully applied to model the effects of climate change and land management practices on watershed yields in past studies. For example, Stone et al. (2001, cited in Cousino et al. 2015) used SWAT to predict that the overall water yield of the Missouri River basin will decrease by 10–20% during spring and summer months but increase during the fall and winter months in response to doubling atmospheric CO_2 concentrations. Cousino et al. (2015) also used SWAT to model the effects of climate change on water, sediment, and nutrient yields from the Maumee River watershed. They reported that moderate climate change scenarios reduced annual flow (up to −24%) and sediment (up to −26%) yields, while a more extreme scenario showed smaller flow reductions (up to −10%) and an increase in sediment (up to +11%).

In this study, the SWAT hydrological model was applied to simulate the impact of climate change on water balance components in the Modder River basin, specifically runoff, water yield, evapotranspiration (ET), and potential evapotranspiration (PET).

2.3.1 Model Setup and Input Data

The first task of setting up SWAT model was to get all the necessary input data in a format required by the model. The input data required by SWAT are the following:

- Digital Elevation Model (DEM): the DEM data with a resolution of 30 m × 30 m was obtained from USGS Earth Explorer. The data was processed using ArcGIS® 10.3.1. This data is key for obtaining the automatic delineation of the watershed by ArcSWAT.
- Land use map in shape file
- Soil map in shape file
- Daily precipitation (of future climate)

- Daily minimum and maximum temperatures (of future climate)

The main aim of running the SWAT hydrological model was to be able to assess the impact of climate change scenarios on water resources, and more specifically on the components of the water balance in the river basin. It should be noted that there were three climate scenarios (historical, RCP4.5 and RCP8.5). The number of combinations of inputs (three climate scenarios; and eight GCMs) will be a total of 24 combinations for each of the precipitation and temperature inputs for this approach.

2.3.2 Model Calibration

The model was calibrated during an earlier study in the upper Modder River basin in a quaternary catchment called C52A (Woyessa et al. 2011; Welderufael et al. 2013). The calibration result was found to be satisfactory according to the Nash–Sutcliffe efficiency value of 0.57 for the monthly streamflow calibration, describing a satisfactory correlation between the observed and simulated monthly stream discharges.

3 Results and Discussion

3.1 Climate Change Modeling

Dynamically downscaled rainfall data, using eight GCM-driven models were analyzed by first establishing the relationships between the historical simulated values (retrospective run) and the measured values. Correlations between the simulated historical values and observed values were established for all the eight GCMs. However, it was found out that the relationships between these variables were not as good as expected for four of them, namely MIROC5, CSIRO-Mk3-6-0, GFDL-EMS2M, and NorESM1-M. Therefore, only the results of those GCMs with strong correlation coefficients between simulated historical and measured values were considered in further analysis and impact assessment. It can be seen that some models predicted more rainfall for all the months of the year while some models predicted less (Fig. 7a). However, when multimodel average is considered, the relationship seems to have improved significantly, as shown in Fig. 7b.

Figures 8a and b illustrate better relationships between the different GCM simulated values and the measured values. Most of the GCMs have a very good fit with the measured values as shown by high R^2 values (0.76 to 0.89). Four of the eight GCMs that performed poorly in predicting the historical rainfall have been excluded from further analysis. The linear relationship between the simulated values and the measured values seems to have improved when the multimodel averages were considered, as shown in Fig. 8b. Given the slightly lower spatial resolution of the GCMs in

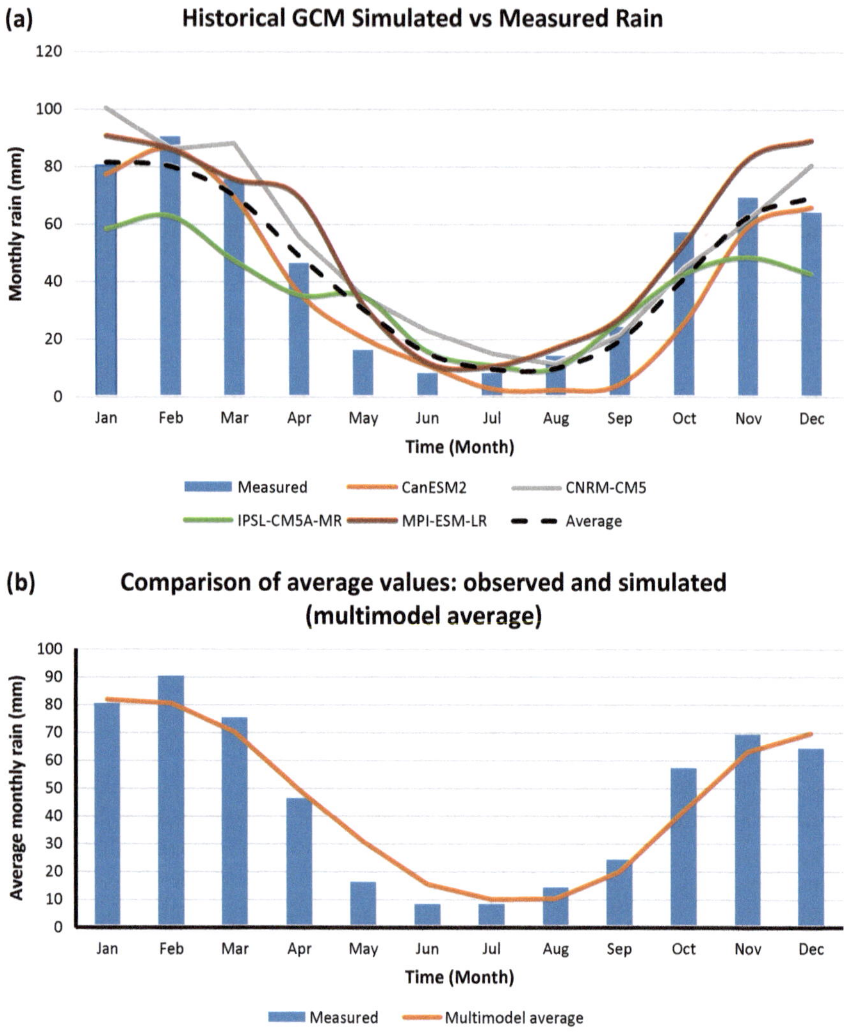

Fig. 7 **a** GCM simulated and measured average monthly rainfall and **b** multimodel monthly average and measured monthly average rainfall

this CORDEX dataset, the prediction capacity of the historical values is quite good, especially when the multimodal averages were considered.

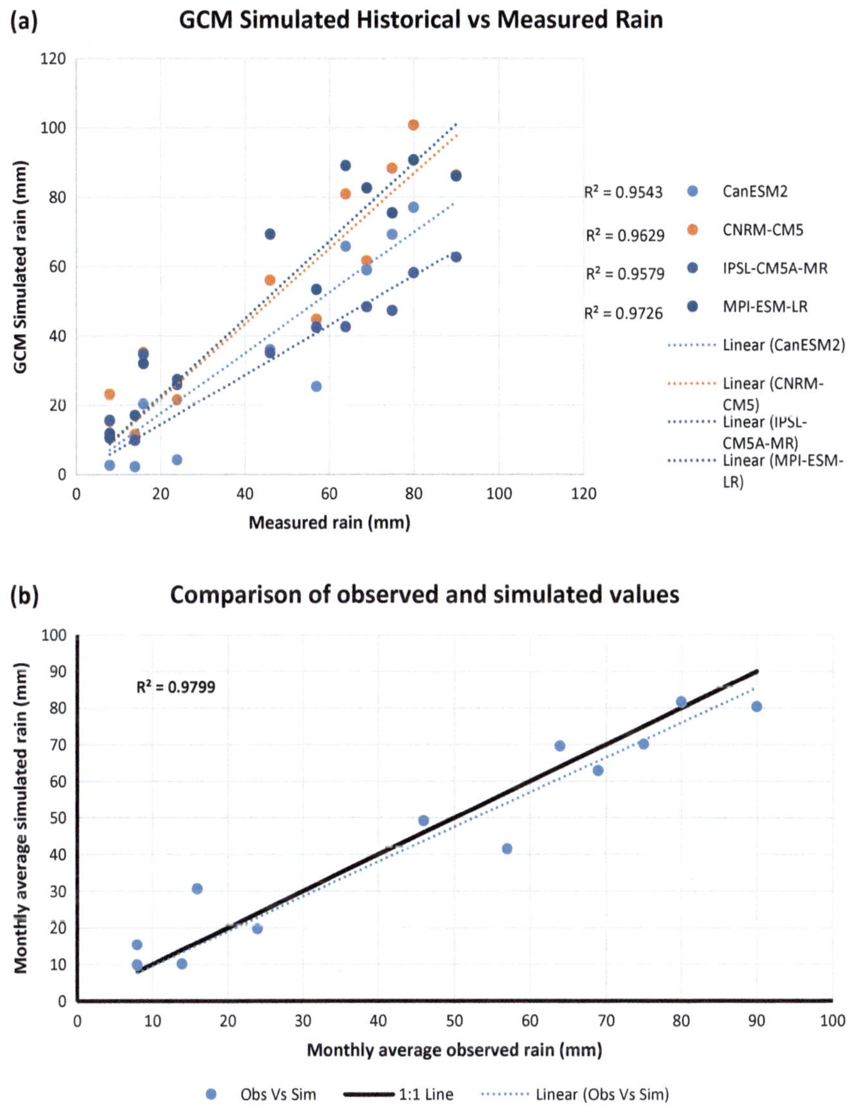

Fig. 8 **a** Observed versus simulated average monthly values of rain using all GCMs and **b** multimodel average

3.2 *Hydrological Modeling*

The results of the SWAT hydrological modeling are presented using climate change data based on CORDEX approach in conjunction with the two greenhouse emission scenarios (RCP4.5 and RCP8.5). In this approach, there were 24 combinations of weather input to a SWAT model, which resulted in the same number of outputs. However, after analysis of the historical simulated rainfall values and comparison with the measured values, four GCMs were excluded from further analysis of the SWAT outputs, which produced monthly summary of the major water balance components of the watershed. The principal factors among them were rain, surface runoff, water yield, evapotranspiration, and potential evapotranspiration. Figure 9a and b presents a comparison of average annual values of rain with the average historical measured values at the two greenhouse emission scenarios. The figures provide an overview of the GCM-driven models output in terms of how the future rain is projected compared to the historical average values. Some of the GCM models predicted a reduction in rainfall while others predicted an increase in rain. The multimodel average, represented by the broken black line (Fig. 9a and b), shows that overall, there will be more monthly rain during some part of the year and less in some months of the year.

Figure 10 summarizes the main water balance components as simulated by SWAT, including all the four GCM models and the three scenarios. In this figure, the potential evapotranspiration showed an increasing trend as it moves from historical through to the RCP8.5 and by all the GCMs. This is a strong reflection of the temperature change scenario, which may indicate potential dry spells between rainy events.

3.3 Impact of Climate Change on Water Resources

In the previous sections, attempts were made to highlight the trends of the future rain and temperature (using PET as a proxy indicator) and the resultant effect on the main components of the watershed water balance, specifically on water yield (flow of water in the river system), evapotranspiration (both actual and potential), as demonstrated using SWAT modeling. In this section, detailed analysis of the impact assessment of climate change will be presented together with the quantification of its impact on water resources.

3.3.1 Impact of Climate Change on Future Rain

In this approach, which considered the future climate for the period of 2020 to 2100, there were differing outcomes in both magnitude and direction of projected change by the different GCMs. Figure 11a and b shows the projected percentage change of average monthly rain by the four GCM models under the two emission scenarios. The comparison was made between the projected values and the historical simulated values. As can be seen from these figures under scenarios of RCP4.5 and RCP8.5, three of the four GCM models projected an increase in rain during the months of May to July and a reduction in rain during the remaining months. The average of the four GCM models is shown in black broken line. This is represented in separate

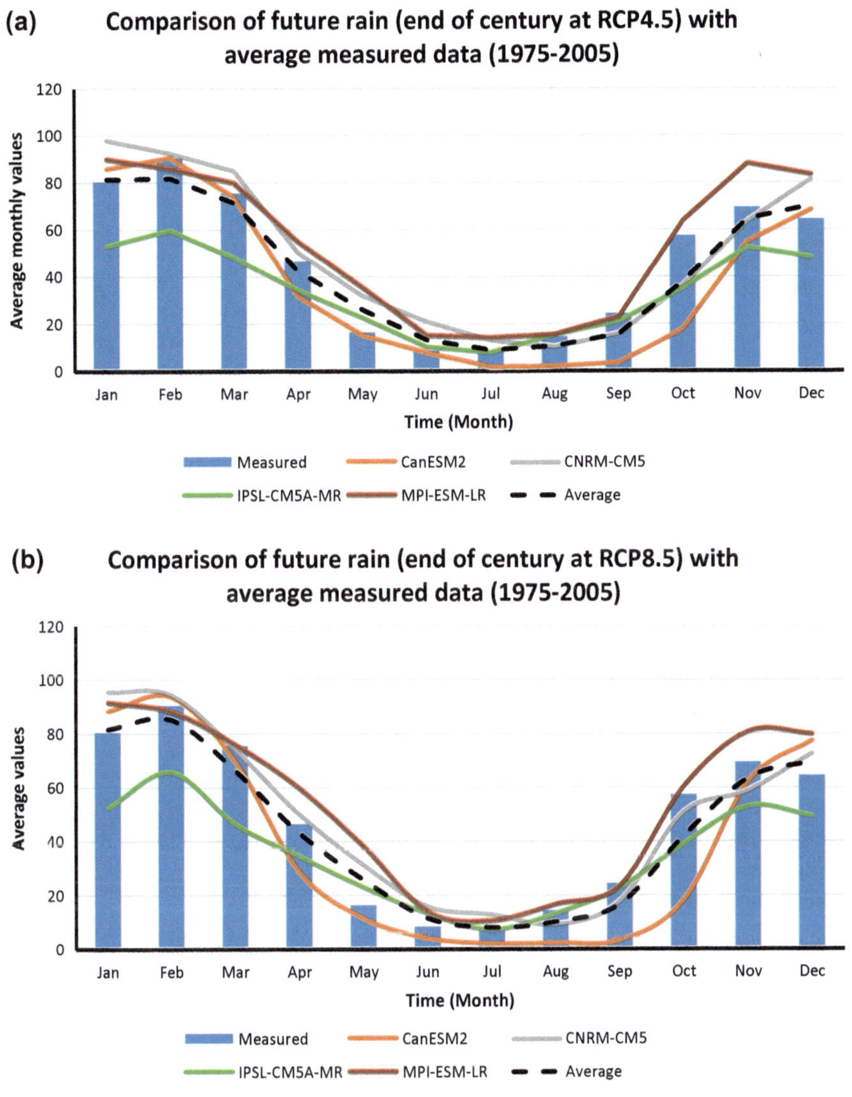

Fig. 9 Comparison of average future monthly rain (end of century) with average measured monthly value: **a** RCP4.5, **b** RCP8.5

figure (Fig. 12) using bar graph for more clarity on the direction and magnitude of the monthly percentage change.

The multimodel average provides a much clearer picture of the percentage change in monthly rain, which is presented in Fig. 12a and b. The result shows that, under the two emission scenarios, the magnitude and direction of percentage change are comparable. Increase in rain is expected in May and June, although these months are not traditionally rainy months for the region. The most important rainy months are

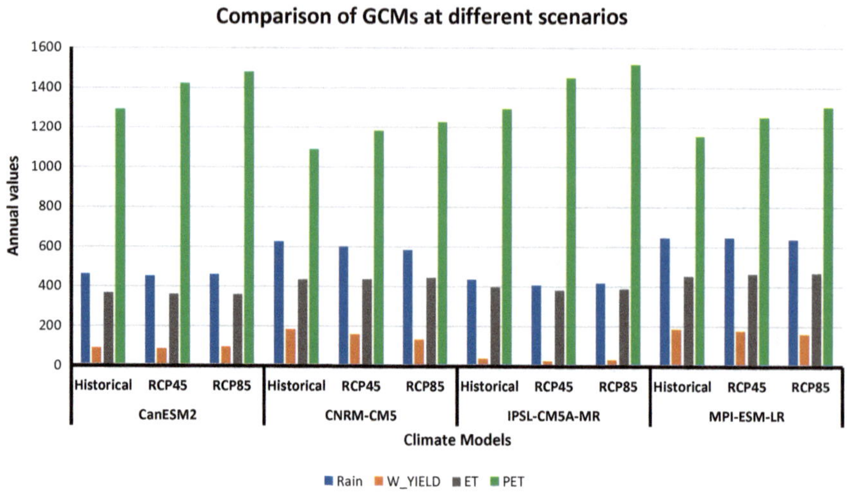

Fig. 10 Average annual values of rain, water yield, ET and PET by the end of the century

September to November and January to March. Any reduction of rain during these months will have a significant impact on water security, particularly on rain-fed agricultural farmers.

Table 2 presents more detailed information and the values of the projected percentage change of future monthly rain for the emission scenario of RCP8.5. The shaded areas in the table show the possible reduction of rain as predicted by three of the four GCM models. However, one of the GCM models (MPI-ESM-LR) showed that there would be zero and/or positive change in the monthly rain for all the months of the year.

3.3.2 Impact of Climate Change on River Basin Water Balance

The projected decrease in rainfall, as described under the previous section, is expected to have an impact on water balance components of the river basin. As shown in the summarized output of the SWAT model using the dynamic downscaling approach, there is a clear indication of an increasing trend of the projected PET by the end of century.

Table 3 shows the projected percentage change of PET compared to simulated historical values at RCP8.5. The percentage change of PET by the end of century (Table 3) shows a significant increase. The multimodel monthly average of percentage change ranges from 17 to 31%. This is reflective of what may be expected by the end of century, where the impact of a higher emission scenario of RCP8.5 on climate change tends to be more clearly detectable.

Increase in PET coupled with a decrease in rain is expected to impact on runoff generation and subsequently on water yield in the river basin. Analysis of this

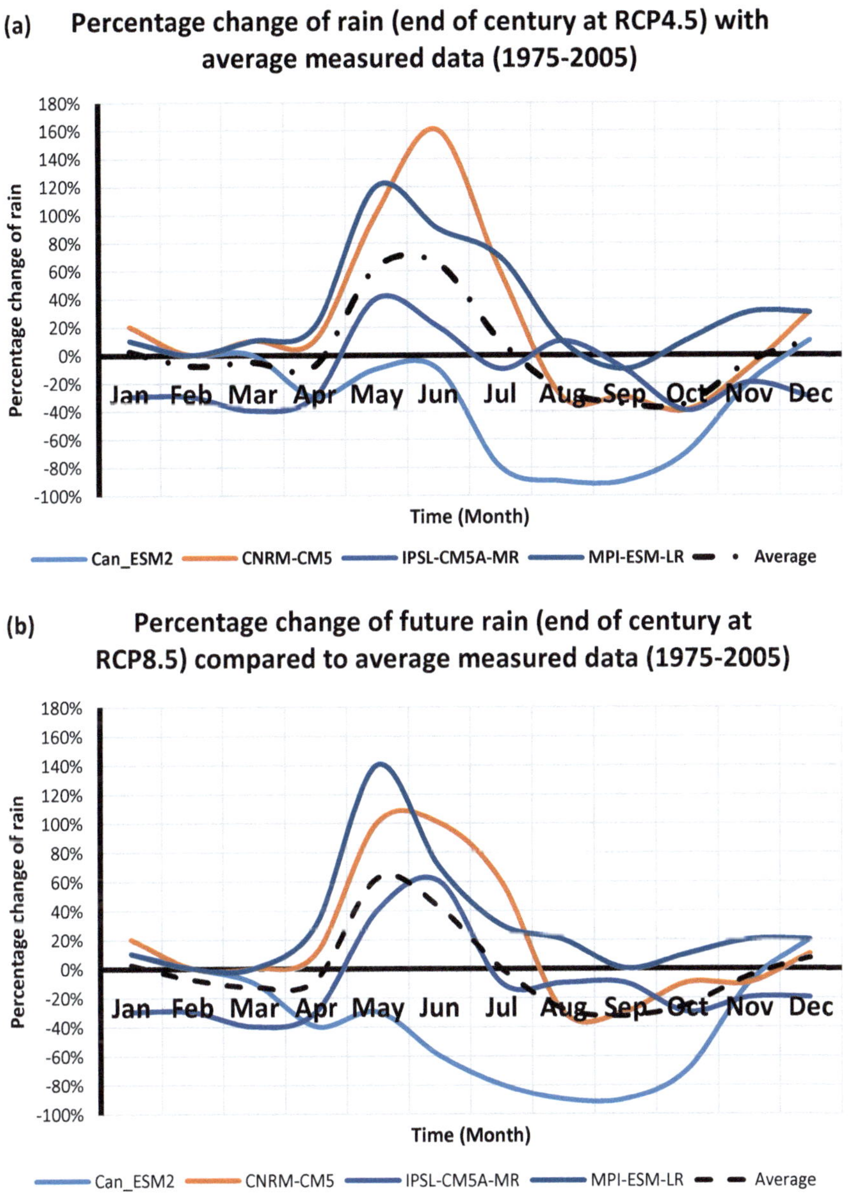

Fig. 11 Percentage change of future rain (end of century) compared to average measured data (1975–2005): **a** RCP4.5, **b** RCP8.5

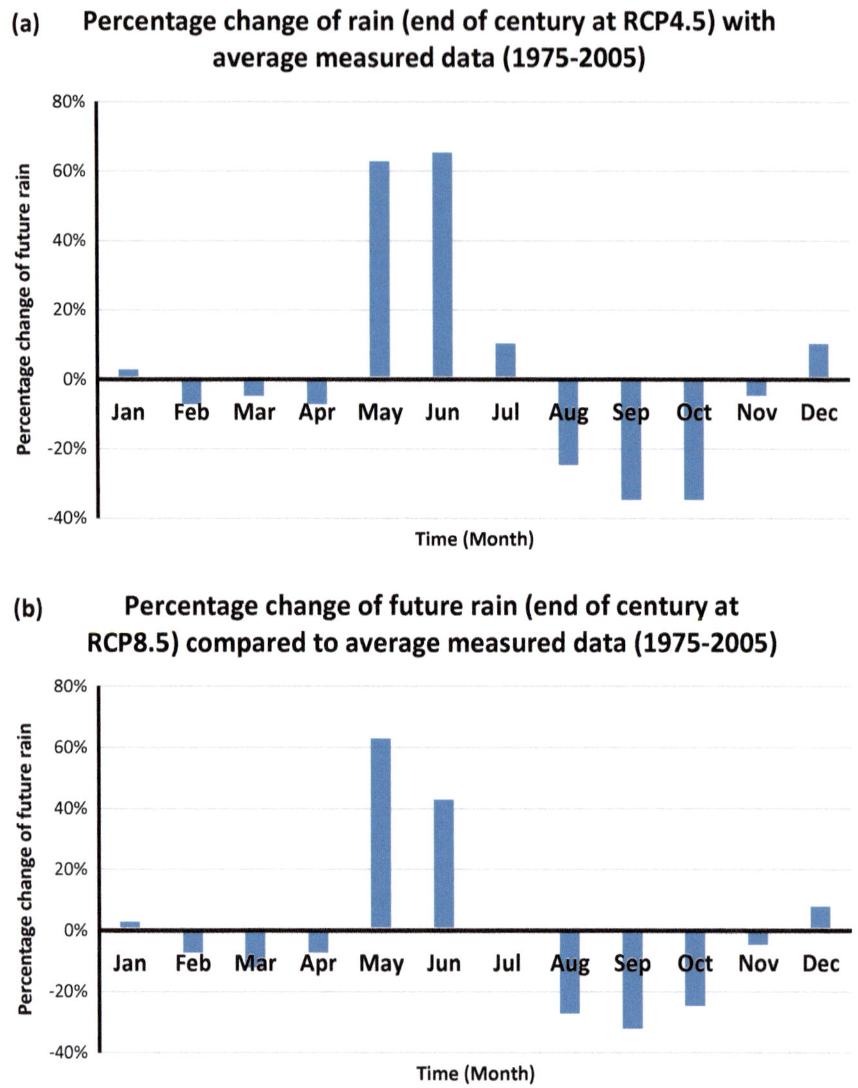

Fig. 12 Multimodel average monthly percentage change of future rain (end of century) compared to average measured data (1975–2005): **a** RCP4.5, **b** RCP8.5

approach using multimodel average shows that there will be a decrease in monthly rain. Similarly, the average water balance components of the watershed, such as rain, surface runoff, and water yield, were simulated using SWAT and analyzed. The results are presented in Fig. 13 showing a decrease in rain by 3%, runoff by 5%, and water yield by 13%.

Table 2 Percentage change of monthly future **rainfall** (end of Century) under scenario of **RCP8.5** compared to average observed values (1975–2005)

Month	Climate Model				Multimodel average (%)
	CanESM2 (%)	CNRM-CM5 (%)	IPSL-CM5A-MR (%)	MPI-ESM-LR (%)	
Jan	10	20	−30	10	3
Feb	0	0	−30	0	−8
Mar	−10	0	−40	0	−13
Apr	−40	10	−30	30	−8
May	−30	100	40	140	63
Jun	−60	100	60	70	43
Jul	−80	60	−10	30	0
Aug	−90	−30	−10	20	−28
Sep	−90	−30	−10	0	−33
Oct	−70	−10	−30	10	−25
Nov	−10	−10	−20	20	−5
Dec	20	10	−20	20	8

Table 3 Percentage change of **PET** (end of century) under scenario of **RCP8.5** compared to historical simulated values

Month	Climate Model				Multimodel average (%)
	CanESM2 (%)	CNRM-CM5 (%)	IPSL-CM5A-MR (%)	MPI-ESM-LR (%)	
Jan	18	2	51	17	22
Feb	12	13	47	18	22
Mar	14	13	53	16	24
Apr	40	16	50	18	31
May	45	9	42	15	28
Jun	38	1	31	18	22
Jul	30	0	23	16	18
Aug	29	8	18	16	18
Sep	34	8	15	11	17
Oct	47	8	28	9	23
Nov	21	15	26	12	18
Dec	18	5	38	12	18

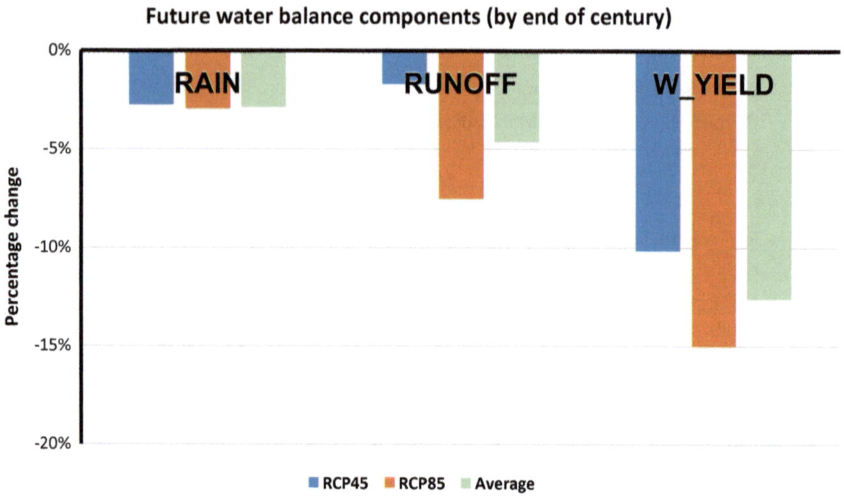

Fig. 13 Future water balance components by the end of the century

4 Water Management Options for Adaptation to Climate Change

Climate change has direct impact on water resources. The management of these resources affects the vulnerability of ecosystems, socio-economic activities, and human health. Climate change is projected to lead to major changes in water availability with increasing water scarcity and droughts, mainly in the central region of South Africa. Water management is, therefore, expected to play an increasingly central role in adaptation to climate change.

In the preceding sections, the impact of climate change on future rain and river basin water balance was highlighted indicating potential decrease in rainfall and possible increase in potential evapotranspiration. In order to illustrate the impact of the climate change scenario on water resources, it is worth considering the case of a reservoir in the Modder river basin. Rusfontein dam is located in the Upper Modder River basin (the upper part of the basin in this study) at the outlet of C52A quaternary catchment. Its contributing area is 973 km^2 and has the full supply capacity of 72.6 million m^3 and full surface area of 11.6 km^2. The main purpose of this dam is to supply water for domestic consumption, industrial use, and irrigation. It was observed that there would be a reduction of an average annual water yield by about 15% (taking the extreme climate change scenario) by the end of the century compared to historical values, which is a reduction from an annual value of 143 mm to 122 mm, a difference of 21 mm. If one considers the contributing area of 937 km^2, there will be a reduction of inflow to the dam by about 20 million m^3 of water annually, which is 28% of the full capacity of the dam. This means that the dam will only be filled to a maximum of 72% of its full capacity. If the scenario of possible increase in PET was

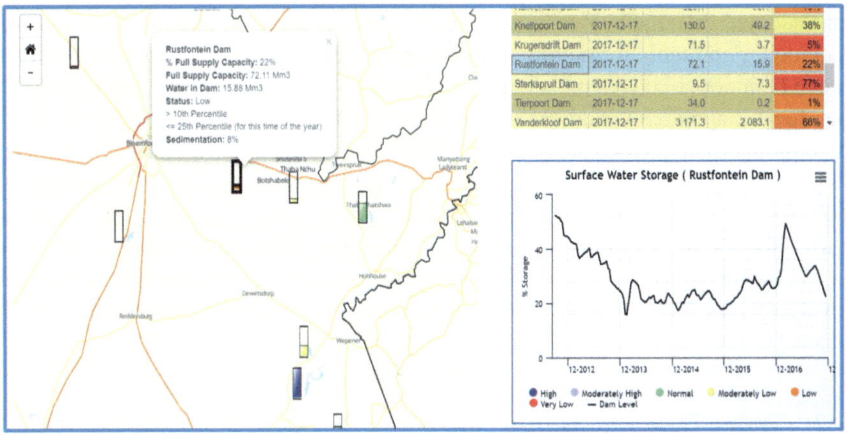

Fig. 14 Interactive map of surface water storage levels in South Africa (*Source* http://niwis.dws.gov.za/niwis2/SurfaceWaterStorage)

considered, there would be much more reduction of water in the reservoir adding to water deficit. In these circumstances, it is important to explore options of water management for adaptation to climate change.

Most parts of South Africa have been experiencing severe droughts in the recent years, particularly from 2013 to 2017. Figure 14 shows the variability of water level in Rustfontein dam from Sep 30, 2012, to Dec 13, 2017. During one of the severe drought months (Dec 2017), Fig. 14 shows that water level in the dam was at a level of 22% of its full capacity, which is very low.

Climate change is expected to intensify the existing risks, particularly in regions where water scarcity is already a concern, such as the central Region of South Africa. In South Africa in general and the central region in particular, agriculture is the main source of food security and rural employment. Any adaptation strategy to climate change must take agriculture at its core as (1) it is the main consumer of freshwater resources, and (2) the rural livelihoods depend on such economic activity.

4.1 Agricultural Water Management

Improving water use efficiency and water productivity in agriculture has been the subject of several studies for many years. However, in Sub-Saharan Africa in general and Southern Africa in particular, there is still a long way to go in this area. The increase in population pressure and the diminishing size of available arable land resources add to the challenge of technological uptake which may lead to an improved water productivity. This is especially true for smallholder and subsistence farmers. One such effort could be aimed at improving rainwater management practices. Rainwater harvesting systems have been developed by researchers (Hensley et al. 2000) and adopted by smallholder farmers in the central region of South Africa, but the techniques and tools for putting such systems in place need to be improved so that it can be more effective and adaptable at a wider scale. Furthermore, there needs to be efficient application and use of water for irrigation under conventional methods by eliminating the losses at various points within the system. Similarly, there needs to be a concerted effort in developing crop varieties and cropping systems that will optimize the use of water for increased water productivity. This means that there is a need for a multidisciplinary approach in addressing the issue of agricultural water management, which requires all the above and stakeholders' involvement at the local level for a potential development of climate-smart agriculture and management tools.

4.2 Water Demand Management

Water demand management is one of the strategies aimed at managing the demand for water when the supply is limited. Brooks (2007) proposes a working definition of water demand management using five components:

1. Reducing the quantity or quality of water required to accomplish a specific task;
2. Adjusting the nature of the task so that it can be accomplished with less water or lower quality water;
3. Reducing losses in movement from source through use to disposal;
4. Shifting time of use to off-peak periods; and
5. Increasing the ability of the system to operate during droughts.

This definition brings out the drivers of water saving and allows for the tracking of gains by the source of the saving.

A national study conducted by WRP Consulting Engineers (McKenzie et al. 2012) indicates that 37% of the total annual water supply in South Africa is considered to be a non-revenue water, which is among the highest when compared to similar water scarce countries such as Australia which has only 10% of non-revenue water. A quarter of this amount is considered to be losses through physical leakage. The current volume of non-revenue water is around 1.58 billion m^3 of water per annum. At a nominal production cost of R4.5/m^3, this loss amounts to about R7.2 billion a year. This represents a huge loss to the economy in general and to the water sector in particular. As indicated above, reduction of water loss is just one of the demand management strategies and it is, therefore, important to manage the demands for water holistically including behavioral change of the users.

4.3 Development and Use of Climate-Smart Water Management Tools

Given the scale and scope of the projected impacts of climate change, it will be essential to identify tools that are effective, implementable, and environmentally beneficial.

4.3.1 Conservation Practices and Precision Agriculture

Precision agriculture, also known as "site-specific crop management," is an information- and technology-based agricultural management system used to identify, analyze, and manage variability within fields for optimum profitability, sustainability, and environmental protection (USDA-NRCS 2006). Agricultural fields often vary in soil types, elevations, soil chemistry, fertility, and productivity. By applying precision agriculture practices, farmers are able to specify the farm input needs, such as nutrient and pesticide application, tillage, and irrigation, throughout an individual field. In addition to cutting production costs and saving energy and water, precision agriculture reduces environmental pollution and improves water quality by reducing nutrient runoff into the water system and the environment (USDA-NRCS 2006). In South Africa, a study by Hendricks (2011) showed that precision agriculture is growing steadily but with slow adoption rate (in spite of the available technology), mainly due to increased cost of imported equipment and technologies. Given its benefit in optimizing the use of agricultural input including water, it is essential to upscale such technology-enabled practice as one of the options in mitigating the impact of climate change over a longer term.

4.3.2 Groundwater Banking

Groundwater banking refers to conjunctive use of surface water and groundwater to improve the reliability of water supply. The general practice is that when surface water is abundantly available, it is used by water users instead of groundwater reserves. However, there needs to be a solution to the question of what happens to the excess of surface water. This is where the concept of groundwater banking was developed which entails that the excess of surface water is pumped into the aquifer artificially recharging the groundwater. The groundwater is then used during dry periods when surface water is less available. The advantage of this system is that it also reduces the loss of water through direct evaporation from surface water storage reservoirs, which is often very high in arid and semi-arid areas. The National Water Resources Strategy of South Africa makes provision for the sustainable use of groundwater resources on a larger scale for sustainable economic growth (DWA 2013). The strategy also emphasizes on conjunctive use of both surface and groundwater. However, it lacks a strategy on the coordinated use of both surface and groundwater and the transfer of one to the other depending on its availability through the groundwater-banking system.

4.3.3 Wastewater Treatment and Reuse

In areas where there is scarcity of conventional water sources (surface water and groundwater), the next available alternative, as one of the adaptation measures to climate change, is to explore the potential of wastewater treatment, recycling, and reuse. These alternative water sources, if treated to an acceptable standard, can be used for non-drinking purposes, such as for agriculture, industrial processes, and other household uses, easing the pressure on conventional water sources. Research (Trinh et al. 2013) has shown that there is a promising potential in the use of treated wastewater for agriculture and at the same time reduction of pollutants, such as nitrogen and phosphorous, from entering the river system and contaminating the environment. In South Africa, interest in the use of treated wastewater for non-drinking purpose was reported to be growing due to decreasing freshwater availability (Adewumi et al. 2010). In most arid areas of South Africa, as freshwater sources become less available due to climate change and population growth, exploring these alternative water sources should be considered as one of the adaptation strategies.

4.3.4 Desalinization

In the preceding sections, several points were highlighted on how best to manage water in order to meet the growing demand against the backdrop of climate change: reduce our use of water (through improved water use efficiency, demand management, and adoption of climate-smart management tools) and recycling wastewater. Another way to address the problem of adequate water supply is to add additional clean water by desalination. Desalination is the process of removing salt from sea or brackish groundwater to make it useable for a range of purposes including drinking. It may thus contribute to adaptation to climate change in all those circumstances in which water scarcity problems may be exacerbated in the future due to climate change and growing water demand.

A serious disadvantage to seawater desalination is, however, the amount of energy it requires. For example, desalination systems, using reverse osmosis technology, require about 30% more energy than the existing inter-basin supply systems delivering water to parts of Southern California (Kelly 2011). However, with the development and adoption of clean energy, such as solar and wind, it has been shown that the amount of energy consumption by desalinization systems has reduced significantly (Gabr 2013). In regions where water is scarce, and the use of the technology is appropriate, intelligent use of desalination technology should be considered and integrated with all other water management options in order to meet the challenge of climate change impact and growing water demand.

5 Conclusion

Investigation of impact of climate change on water resources was conducted in the Modder River basin (central region of South Africa). In this study, Dynamical Downscaling approach (CORDEX) was adopted. The CORDEX dataset was used with eight GCM models. For each of the GCM models, two greenhouse gas emission scenarios, namely RCP4.5 and RCP8.5, were considered. The weather data obtained from the climate change scenarios were entered into SWAT hydrological model to simulate the impact of climate change on water and river basin water balance components for the end of century period.

Out of the eight GCMs considered in this dataset, four of them produced weak correlation between measured and simulated historical rainfall values (retrospective runs), hence not considered for impact analysis. However, the multimodel average of the four GCMs gave a strong correlation between measured and simulated historical rainfall values ($R^2 = 0.92$) and was used in the analysis of the impact of climate change on rainfall and river basin water balance components. The hydrological simulation showed that there was a steady increase in PET with the two emission scenarios when compared to historical values. However, the impact on rainfall was not unidirectional, i.e., there was an expected increase during the normally dry months of the year (May to July) while the rainfall is expected to decrease during the normal rainy months (September to November and February to March). On average, the study revealed the following as possible impact of climate change by the end of the century: Rainfall is expected to decrease by 3%; Surface runoff is expected to decrease by 5%; Water yield is expected to decrease by 13%; and PET will increase by 22%. The expected decrease in rainfall coupled with increase in PET are going to be the two major drivers in exacerbating the existing problem of water scarcity in the region.

References

Adewumi JR, Ilemobade AA, Van Zyl JE (2010) (2010) Treated wastewater reuse in South Africa: overview, potential and challenges. Resour Conserv Recycl 55(2):221–231

African Ministers' Council on Water (AMCOW) (2012) Water sSecurity and climate resilient development: technical background document. african ministers' council on water

Arnold JG, Srinivasan R, Muttiah RS, Williams JR (1998) Large-area hydrologic modeling and assessment: Part I. Model development. J Am Water Resour Assoc 34(1):73–89

Brooks DB (2007) An operational definition of water demand management. Int J Water Resour Dev 22(4)

Cousino LK, Becker RH, Zmijewski KA (2015) Modeling the effects of climate change on water, sediment, and nutrient yields from the Maumee River watershed. J Hydrol: RegNal Stud 4:762–775

Department of Environmetal Affairs (2017) South African national land cover change assessment dataset. https://egis.environment.gov.za/data_egis/data_download/current

Department of Water Affairs (2013) National water resources strategy, Pretoria, South Africa.http://www.dwa.gov.za/documents/Other/Strategic%20Plan/NWRS2-Final-email-version.pdf

Free State Provincial Government (n.d.). Overview of the provincial revenue and expenditure
Gabr SM (2013) Solar desalination as an adaptation tool for climate change impacts on the water resources of Egypt. UNESCO, Technical Report
Gassman PW, Reyes MR, Green CH, Arnold JG (2007) The soil and water assessment tool: historical development, applications, and future research directions. Trans ASABE 50(4):1211–1250
Giorgi F, Jones C, Asrar GR (2009) Addressing climate information needs at the regional level: the CORDEX framework. WMO Bull 58(3):175–183
Hendriks J (2011) An analysis of precision agriculture in the South African summer grain producing areas, Mini-dissertation, the North-West University. South Africa
Hensley M, Botha JJ, Anderson JJ, van Staden PP, du Toit A (2000) Optimizing rainfall use efficiency for developing farmers with limited access to irrigation water. WRC Report No. 878/1/00 Water Research Commission, Pretoria
Jain MK, Kothyari UC, Raju KGR (2004) A GIS based distributed rainfall – runoff model. J Hydrol 299:107–135
Jewitt GPW, Garratt JA, Calder IR, Fuller L (2004) Water resource planning and modelling tools for the assessment of land use change in the Luvuvhu Catchment, South Africa. Phys Chem Earth 29:1233–1241
Kelley AH (2011) Seawater desalination: climate change adaptation strategy or contributor? Angela Haren Kelley
Kundhlande G, Groenewald DC, Baipethi MN, Viljoen MF, Botha JJ, van Rensburg LD, Anderson JJ (2004) Socio-economic study on water conservation techniques in semi-arid areas. WRC Report No. 1267/1/04, Water Research Commission, Pretoria
Levina E (2006) Domestic policy frameworks for adaptation to climate change in the water sector Part II: Non-annex i countries. Lessons learned from Mexico, India, Argentina and Zimbabwe, OECD, Paris
MacKellar N, New M, Jack C (2014) Observed and modelled trends in rainfall and temperature for South Africa: 1960–2010. S Afr J Sci 110(7–8):1–13
Mckenzie R, Siqalaba ZN, Wegelin WA (2012) The state of non-revenue water in South Africa. WRC Report No. TT 522/12, Water Research Commission, Pretoria
Mangaung Local Municipality (MLM) 2015/16. Reviewed integrated development plan. Mangaung Local Municipality
Peugeot C, Esteves M, Galle S, Rajot JL, Vandervaere JP (1997) Runoff generation processes: results and analysis of field data collected at the East Central Super site of the HAPEX-Sahel experiment. J Hydrol 188–189:179–202
Puukka J, Dubarle P, McKiernan H, Reddy J, Wade P (2012) Higher education in regional and city development. The Free State, South Africa. OECD, Paris
Trinh L, Duong CC, Van Der Steen P, Lens PNL (2013) Exploring the potential for wastewater reuse in agriculture as a climate change adaptation measure for Can Tho City. Vietnam Agric Water Manag 128:43–54
Turpie JK, Marais C, Blignaut JN (2008) The working for water programme: Evolution of a payments for ecosystem services mechanism that addresses both poverty and ecosystem service delivery in South Africa. Ecol Econ 65(4):788–798
USDA-NRCS (2006) Conservation practices that save: precision agriculture. https://www.nrcs.usda.gov/Internet/FSE_DOCUMENTS/nrcs143_023626.pdf
United States Geological Survey (USGS) Earth Explorer (2017) Digital Elevation Model. https://earthexplorer.usgs.gov. Accessed 7 Oct 2017

Welderufael WA, Woyessa YE, Edossa DC (2013) Impact of rainwater harvesting on water resources of the modder river basin, central region of South Africa. J Agri Water Manag 116:218–227

World Water Council (2000) World water vision: making water everybody's business. Earthscan Publications Ltd., London

Woyessa YE, Welderufael WA, Kinyua JDM, Kundhlande G, Tsumake OP (2011) Land-water linkages: agent based modelling of land use change and its impact on the water resources. WRC report no. 1753/1/10; ISBN 978-1-4312-0083-2

Woyessa YE, Pretorius E, van Heerden PS, Hensley M, van Rensburg LD (2006) Impact of land use on river basin water balance: a case study of the Modder River basin, South Africa, Colombo, Sri Lanka: Comprehensive Assessment Secretariat. Comprehensive Assessment Research Report 12

Behaviour Theories to Understand Perception of Public Towards Acceptance of Recycled Wastewater

Samiya Gul, Ivan Govender, and Faizal Bux

1 Introduction

Water is a finite and most valuable resource. The distribution of water on the planet is approximately 98% salty water and 2%is fresh water. Out of the 2%, almost 70% is snow and ice, 30% is groundwater, less than 0.5% is surface water (lakes, rivers, etc.) and less than 0.05% is in the atmosphere. This is where the problem of fresh water scarcity has begun.

Climate change has also posed some serious problems of water scarcity like the most alarming concerns of fluctuating precipitation, increased global mean surface temperature, sea level rise, and extreme weather conditions like cyclones, droughts, and floods. Climate change affects the hydrological cycle and water availability in the way that global warming causes polar ice to melt into the sea and that turns fresh water into sea water. The melting of glaciers will increase water supply to rivers and lakes, but only for short to medium term. In the dry regions, climate change leads to reduced rainfalls but the overall effect is the extreme floods and droughts globally.

Consequences of the changing patterns of the hydrological cycle, water availability, water demand, water quality, and water security have been violated. Water is an important medium that transfers the effects of climate change to whole of the environment including human beings.

The different possible solutions against the climate change effects on the accessibility to potable water are, redistribution of fresh water over space and time, reservoirs to store the fresh water, and pipelines to distribute it. The most important can

S. Gul · I. Govender
Department of Entrepreneurial Studies and Management, Durban University of Technology, PO Box 1334, Durban 4000, South Africa

S. Gul · F. Bux (✉)
Institute for Water and Wastewater Technology, Durban University of Technology, PO Box 1334, Durban 4000, South Africa
e-mail: faizalb@dut.ac.za

Y. Li et al. (eds.), *BRICS Countries: Sustainable Water Resource Management and Pollution Control*, https://doi.org/10.1007/978-981-99-9581-3_9

be desalination and reuse and recycling of waste water. Also, efforts must be made to increase the benefits of water saving and the various options to do so among the common masses. One of the important solutions to handle this problem is wastewater reuse, which has been considered as an important part of Sustainable Water Management Scheme (Marlow et al. 2013).

2 Wastewater Reuse as a Solution to Water Scarcity

The universal issues that communities face around the world is water scarcity. According to a recent study conducted by the American Association for the Advancement of Science, 71% of the world's population experience drought every year and with half a billion people face severe water scarcity. With the increase in the population, there will be a greater demand for ~~on~~ fresh and clean water. So, securing water is not an option but a priority. According to the recent report of UN, about half of the global population will be facing water shortages by 2030 and it is forecasted that demand will be approximately 40% greater than supply (WWAP 2015). In the twenty-first century, one of the primary challenges facing many countries is water scarcity. It is affecting every continent. Almost one-fifth of the world's population is living under the stress of physical scarcity and 500 million people are predicted to approach this condition. Almost one-quarter of the world's population has economic water shortages (UNDP 2006).

Water reuse has been identified as an effective strategy to resolve the water scarcity problem. The level of interest and acceptance is likely to increase because of two basic advantages; mitigating the environmental impact of wastewater disposal and creating an additional water supply source. Water reuse is economically friendly and innovative that produces high-quality water at lower cost than developing a new water supply. The countries that are successfully adapting the water reuse technologies are California, Israel, and Singapore (Joe 2016). Wastewater reuse is known as a key strategy in the conservation of water resources and solution for global water challenges (Wester et al. 2015). Governmental support, efficient technology, sufficient capital, and public acceptance are among essential requirements for the success of wastewater reuse. Negative responses from public are among the main hurdle for successful implementation of these wastewater reuse schemes (Nancarrow et al. 2008; 2009; Khan and Gerrard 2006; Chen et al. 2012; Baghapour et al. 2017).

3 Factors Influencing Public Perception of Reclaimed Water

The perception of wastewater reuse may be influenced by several factors such as emotions, risk, trust in service provider, and cost of produced water, which may influence the decision-making processes. Hence, there is a need to identify key factors, which can influence the public perception and decision-making of accepting or rejecting water reuse. Factors which can influence the perception of wastewater reuse are summarized below.

3.1 Emotions/Disgust/Yuck Factor

Since the early 1970s, the "yuck factor" or "disgust" factor can be regarded as a severe challenge for accepting recycled water. The public considers the use of recycled water to be a psychological barrier, which appears to be the disgust emotion, resulting from the thought of reusing wastewater. Even some participants are willing to use the recycled water for all purposes, including drinking water but at the same time, they also admitted that they would prefer to drink bottled water or pass their household drinking water through a filter (Kaercher et al. 2003). The public had associated recycled water with waste and portrayed a mental image of "disgust". The "disgust emotion" refers to emotional discomfort that is generated from close contact with some unpleasant stimuli. Thus, a disgust reaction is generated as people perceive recycled water as unclean or unpleasant. Therefore, current projects avoided terms such as "treated waste water". For example, recycled water is termed "NEWater" in Singapore and "Repurified water" in San Diego (Leovy 1997).

3.2 Risk Factor

The perception of risks with recycled water plays an important role in deciding water reuse for the public (Hartley 2006). The primary concern of recycled water is the associated health risk after consumption. The level of public acceptance may increase if the water quality complies with international standards (Hartley 2006; Rivett et al. 2013). The influence of the risk factors on public acceptance of reclaimed water was reported in literature. A survey by Sydney Water (1999) regarding a water reuse scheme, found that 59% of respondents agreed with the statement that "no one can guarantee the safety of recycled water". Some studies reported significant relationships between risk perceptions and trust in the institutions and organizations. Studies have shown higher levels of trust in relation to lower perceptions of risk and that in turn can increase public acceptance (Nancarrow et al. 2009; Ross et al. 2014).

Therefore, suggesting that acceptance level, risk, and trust should have an indirect relation.

Eiser et al. (2002), reported two ways which may link trust and risk perceptions (viz. causal-chain and associationist view). According to causal-chain view, factors such as trust and risk are inter-related for any decision-making, in other words, trust depends upon how people process and interpret risk-related information about the technology. This may assist in shaping the decision to accept or decline the technology. Whereas, an associationist view holds that both trust and risk perceptions are independent, meaning the acceptance of a technology can be determined by both underlying attitudes of trust and risk perceptions. The associationist model was strongly supported in contrast to the causal model. Various studies have found that trust is a consequence of attitudes rather than the driver for acceptance (Frewer et al. 2003). Moreover, in the case of water reuse, the associationist model plays a significant role, as the trust and risk perceptions play individual roles. Therefore, trust, risk, and the emotional reaction can be strong predictors in deciding the behaviour of public for water reuse.

3.3 Trust in Service Provider

Trust can be described as a multidimensional and complex construct and a psychological state that includes the intention to accept vulnerability based upon positive expectations on the behaviour of the authority responsible for recycled water schemes (Fischhoff 1999; Siegrist et al. 2000). According to Siegrist et al. (2000), many individuals make decisions based on scientific research; however, they lack resources such as knowledge, time, and interest. Therefore, they rely on trust from authorities and government agencies to make decisions. Hurlimann et al. (2008) investigated the role of trust in public acceptance on potable water reuse. The past research and theory also provides a strong basis for the hypothesis that higher trust in water authority to deliver safe recycled water will be related to the lower risk perceptions and thus leading to the greater acceptance of the recycled water. Ross et al. (2014) studied the role of trust in authorities in relation to perceived procedural fairness, social identity, source credibility, and the acceptance of recycled water management schemes.

3.4 Social Set up

Social and geographic factors influence the acceptance of recycled water. However, proper justifications based on these observations should be further supported while concluding and generalizing results from surveys (Kaercher et al. 2003). Acceptability for water reuse has shown a high correlation with men and those with higher levels of education (Dolnicar and Hurlimann 2010). In other studies, the age of participants was also related to public acceptance (Dolnicar and Hurlimann 2010;

Dolnicar et al. 2010). Some studies had reported no significant relationships between recycled water acceptance and socio-demographics. Jeffrey and Jefferson (2003) further reported no significant relationship between acceptance of water reuse and socio-demographics in the United Kingdom. Hence, the socio-demographics has an important role to select a representative sample for the survey.

3.5 End Use

The final usage of recycled water can impact public acceptance. Acceptance is generally higher when there is minimal human contact (Hartley 2006), which further led to a higher acceptance of using recycled water for industrial applications (Kaercher et al. 2003). This was further supported by Dolnicar et al. (2010), whereby 92% of respondents agreed that they would use recycled water for irrigation, however, only 36% confirmed for drinking water.

3.6 Cost

The cost factor can increase or decrease public acceptance of reclaimed water. Cheaper drinking water sources can increase their willingness to purchase and consumption. No change in tariff can shift public emotions to other factors such as disgust leading to lower acceptance of recycled water. Hartley (2006) reported that public acceptance of water reuse is higher when costs of network installation and running are reasonable. However, the cost factor depends on various parameters and its value is attached to capital investments made by service providers and returns (Woolston and Jaffer 2005).

4 Theories of Individual Behaviour and Behaviour Change

Theories and models are the structures by which human behaviour can be conceptualized. The theories almost originate from all social science subjects and are used by researchers in various contexts and situations. Developments are being done on already existing theories to customize their application beyond the concepts of psychology to the real-life issues. In behaviour theories, the locus of behaviour is the individual, and impact of various external factors on the behaviour is conceptualized. A number of theories are available that study the behaviour of individual towards an external factor, however the theories that are relevant to the social aspects of recycled wastewater are summarized in next sections.

4.1 *The Theory of Planned Behaviour and Theory of Reasoned Action*

One of the most influential social-psychological models for describing human behaviour is theory of planned behaviour (TPB) (Armitage and Conner 2001). This theory is considered as the most effective tool in predicting human behaviour (Ogden 2003). According to the theory of reasoned action, intention is a key factor in determining behaviour. Intention presents itself as the best predictor of behaviour and is the consequence of the combination of attitudes towards a behaviour. Individual's intention is the most important factor to perform behaviour (Han and Stoel 2017). The positive or negative intentions from personal attitudes and social pressures (subjective norms) shape the desired behaviour (Ajzen and Fishbein 1975).

Ajzen in 1985 expanded upon the theory of reasoned action and formulated the theory of planned behaviour. According to this theory, the individual attitudes and beliefs explain the behaviour. This theory added a third set of factors affecting intention or perceived behavioural control. This is similar to self-efficacy which means the perceived ease or difficulty with which the individual will be able to achieve or carry out the behaviour. The foundational hypothesis of this theory is that behaviour is directly determined by the intention which in turn is explained by the attitude towards the behaviour, the subjective norms, and perceived behavioural control (Ajzen 1991). Attitude is the sum of all behavioural beliefs, the expectation that showing a behaviour would result in a particular outcome. Subjective norms in other words are the social pressure, are the perceived expectations what other people perform the alternative behaviour times the willingness to perform that expectation. It is the influence of other significant factors to perform the behaviour. Perceived behavioural control is a degree people have for the opportunity and ability to do a particular alternate behaviour. It is the ease or difficulty of performing the behaviour (see Fig. 1).

Strong correlations were found between behaviour and both the attitudes and perceived behavioural control components. This theory is used to explain and predict the likely behaviour but is not considered as useful in planning and designing the

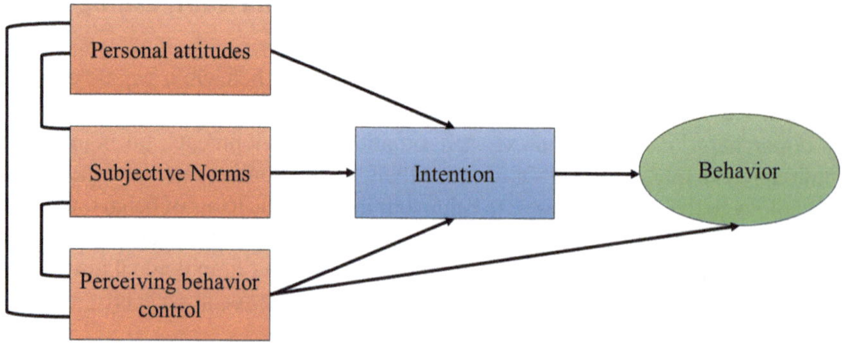

Fig. 1 Construct of the theory of planned behaviour (Ajzen 1991, p. 182)

type of intervention for behaviour (Taylor et al. 2007; Webb et al. 2010). This theory was widely applied in diverse sectors; software copying, ethical purchase behaviour, purchase of fair-trade products, and avoidance behaviour towards counterfeit products (Han and Stoel 2017). However, the TPB theory was also criticized in terms of predictability. Ogden (2003) found that the TPB theory showed no effect of attitude, subjective norms, or perceived behavioural control on behaviour. Ajzen and Fishbein (2004) clarified that the comparative importance of these three constructs may vary depending on the type of behaviour and population. Additionally, Ajzen (1991) showed that the TPB theory may allow the addition of other predictors (other than these three factors) if they increase the predictive power of the model. Ajzen (1991), further suggested that this model is open to the inclusion of additional predictors. With this inclusion, it has been widely used and applied in different sectors.

4.2 *The Norm-Activation Theory*

The Norm-Activation Theory (NAT) is a conventional behaviour theory, which is applied to study specifically the helping behaviour of human beings. NAT was applied to several environmental situations involving public psychology and behaviour. The theory is based on the construct proposed by Schwartz and Howard (1981) by which they were mainly interested to investigate the root reasons, which can motivate an individual to help others. The theory is conjectured on the assumption that an individual may help others only if he feels morally obliged towards it. This moral obligation was coined in this theory as activated personal norm (Schwartz and Howard 1981). The active personal norm is a reflection of identity of an individual and needs to be activated before expecting a particular behaviour from an individual. Schwartz and Howard (1981) proposed four criteria which need to be fulfilled to activate the personal norm in an individual; (1) The person should be made aware of the need for help, which is reflected as awareness of need. (2) The person should be knowledgeable about the consequences of helping, an attribute that is referred to as awareness of consequences in the theory. (3) The person needs to accept the personal responsibility of helping and this attribute is referred to as acknowledgement of the responsibility; and (4) the person should recognize that he or she is capable of controlling his/her behaviour, which is referred as perceived behavioural control in the theory structure. The theory blends awareness of need, consequences and the personal behaviour control of an individual empirically into a construct as illustrated in Fig. 2 (Klockner 2013).

NAT has been applied to study pro-environmental behaviour (Hunecke et al. 2001; Harland et al. 2007). Thogersen (1996) argued that NAT is more suitable than TPB to study the environmental behaviour because the variables in former theory are more relevant to moral behaviour which is a pre-requisite in most of the environmental behaviour situations. The variables in TBP are more measurable of cost–benefit factors which are non-moral and less relevant in the exhibition of environmental behaviour.

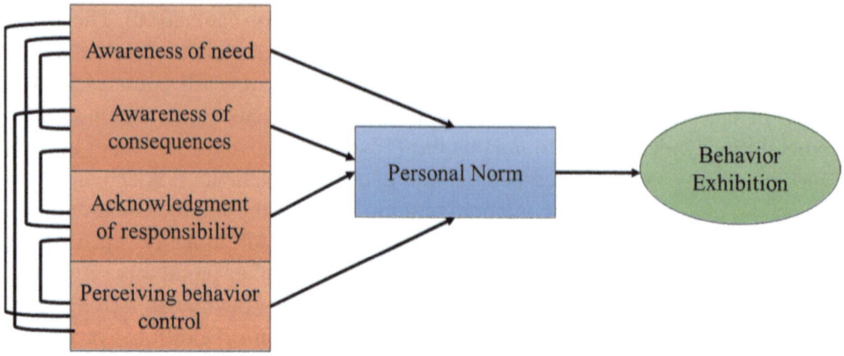

Fig. 2 Illustration of the construct of the NAT

In recycling awareness of the need for recycled water can be perceived by awareness of the scarcity of the water in a particular region and history or prediction of droughts. The consequences of non-availability of fresh water and the benefits of wastewater recycling can be referred to as awareness of consequences. Personal responsibility towards accepting recycled water is a complex attribute in behaviour control towards wastewater recycling. The recycled water is a purchased item and most often the consumer is having the authoritative attribute. However, the responsibility can be triggered in a broader context of being a citizen (or community) in the water scarcity country, which made it everyone's responsibility towards other people, who cannot afford expensive bottled water.

4.3 Value Belief Norm Theory

The value belief norm (VBN) theory was proposed by Stern in 2000 (Stern 2000). Likewise NAT, it assumes that the behaviour is a consequence of the personal norms which are driven by awareness of need, consequences, and responsibility. However, VBN assumes that environmental beliefs are also linked to personal norms and makes an integrative theory in itself. A sequence of variables from personal values and awareness to the environmental concerns of human activity are suggested in this theory to activate sustainable personal norms in an individual. Awareness of consequences in VBN is related to ecological setup of the nature which is referred to New Environmental Paradigm (NEP) in this theory. This attribute of NEP refers to the acceptance of conventional beliefs pertaining to disturbance in the environment due to human activity such as polluting fresh water sources, over consumption of natural resources, limitation of non-renewable energy resources, etc. The function of NEP in VBN theory is to create a linkage between personal values and personal norms, rendering it more applicable in environmental behaviour studies. Application

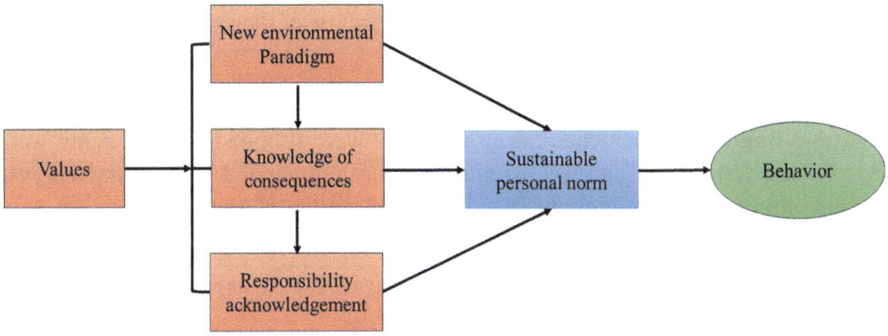

Fig. 3 Illustration of the construct of the VBN theory, Klockner (2013)

of VBN theory in environmental behaviour research has been reported by Poortinga et al. (2004), De Groot and Steg (2007), Hansla et al. (2008) (see Fig. 3).

5 Application of Behaviour Theories in Wastewater Reuse

Ajzen's theory of planned behaviour was used to predict the intended behaviour for recycled wastewater. Nancarrow et al. (2008) proposed a predictive model based on Ajzen's model. However, some additional factors were included from comparable literatures to predict the behaviour. The factors used by the authors are as follows:

- Emotion: it can be described as "yuck factor" in terms of wastewater. It is the degree, to which a person feels negative or positive emotions towards the recycled wastewater.
- Attitudes: the degree to which a person believes that accepting and supporting the recycled water will bring positive outcomes.
- Subjective norms: it is the social pressure, that a person feels from other people to support this behaviour.
- Perceived control: it is the degree of control, a person feels they have over the source and quality of water.
- Risk perceptions: the level of risk perceived by a person linked with the recycling of wastewater.
- Knowledge: the amount of knowledge a person has about the issues and benefits of recycled wastewater.
- Trust: the level to which a person trusts the authorities and the people involved in implementing and managing the recycled wastewater.
- Responsibility: the level to which a person rates the relative responsibility of the community and the authorities for making sure there is enough water for the future.
- Environmental obligation: the level to which a person feels personally obligated to protect the environment.

- Intended behaviour: the intention to behave in a way that accepts or rejects the recycled wastewater.

These variables were tested in the model through questionnaires. The model underwent retesting and refinements. Some variables were dropped out and some were added. Knowledge was found as insignificant for the intended behaviour. Risk was divided into two separate components; health risk and systemic risk.

6 Conclusions

Reclaimed water is a promising strategy as an emergency water source to deal with water scarcity. The technological advancements made it possible to reduce the health risk associated with reclaimed water. However, the social aspects are still a challenge to the successful implementation of reclaimed water. Behavioural models are one of the strategies to understand the key factors responsible for the public rejection of reclaimed water. This chapter summarizes the various factors, which are responsible for public reluctance, and documents the behaviour models that can be applied to understand the public behaviour towards reclaimed water. To date, only TPB has been applied to predict the intentions of the public. However, there is a potential for other theories such as the NAT and VBN theory which can be useful to understand the public behaviour towards the reclaimed water. These theories have been discussed in relevance to the reuse of wastewater and future recommendations are to investigate the potential use of these behavioural models to develop an in-depth understanding of the social aspects of wastewater reuse.

References

Ajzen I (1991) The theory of planned behavior. Organ Behav Hum Decis Process 50:179–211

Ajzen I, Fishbein M (1975) A Bayesian analysis of attribution processes. Psychol Bull 82(2):261

Ajzen I, Fishbein M (2004) Questions raised by a reasoned action approach: comment on Ogden (2003)

Armitage CJ, Conner M (2001) Efficacy of the theory of planned behaviour: a meta-analytic review. Br J Soc Psychol 40(4):471–499

Baghapour MA, Shooshtarian MR, Djahed B (2017) A survey of attitudes and acceptance of wastewater reuse in Iran: Shiraz city as a case study. Journal of Water Reuse and Desalination 7(4):511–519

Chen Z, Ngo HH, Guo WS (2012) A critical review on sustainability assessment of recycled water schemes. Sci Total Environ 426:13–31

De Groot J, Steg L (2007) General beliefs and the theory of planned behavior: the role of environmental concerns in the TPB. J Appl Soc Psychol 37:1817–1836

Dolnicar S, Hurlimann A (2010) Drinking water from alternative water sources: differences in beliefs, social norms and factors of perceived behavioral control across eight Australian locations. Water Sci Technol 60(6):1433–1444

Dolnicar S, Hurlimann A, Nghiem LD (2010) The effect of information on public acceptance-the case of water from alternative sources. J Environ Manag 91:1288–1293
Dolnicar S, Schäfer AI (2009) Desalinated versus recycled water: public perceptions and profiles of the accepters. J Environ Manag 90:888–900
Eiser JR, Miles S, Frewer LJ (2002) Trust, perceived risk and attitudes towards food technologies. J Appl Soc Psychol 32:2423–2433
Fischhoff B (1999) If trust is so good why isn't there more of it? In: Cvetkovich G, Lofstedt R (eds) Social trust and the management of risk. Earthscan, London
Frewer LJ, Scholderer J, Bredahl L (2003) Communicating about the risks and benefits of genetically modified foods: the mediating role of trust. Risk Anal 23(6):1117–1133
Hartley TW (2006) Public perception and participation in water reuse. Desalination 187:115–126
Han TI, Stoel L (2017) Explaining socially responsible consumer behavior: a meta-analytic review of theory of planned behavior. J Int Consum Mark 29(2):91–103
Hansla AH, Gamble A, Juliusson A, Gärling T (2008) Psychological determinants of attitudes towards and willingness to pay for green electricity. Energy Policy 36:768–774
Harland P, Staats H, Wilke AM (2007) Situational and personality factors as direct or personal norm mediated predictors of pro-environmental behavior: questions derived from norm-activation theory. Basic Appl Soc Psychol 29:323–334
Hunecke M, Blöbaum A, Matthies E, Höger R (2001) Responsibility and environment: ecological norm orientation and external factors in the domain of travel mode choice behavior. Environ Behav 33:830–852
Hurlimann AC, Hemphill E, Mckay J, Geursen G (2008) Establishing components of community satisfaction with recycled water use through a structural equation model. J Environ Manag 88(4):1221–1232
Jeffrey P, Jefferson B (2003) Public receptivity regarding "in-house" water recycling: results from a UK survey. Water Sci Technol: Water Supply 3(3):109–116
Joe V (2016) Water reuse: a critical step on the path to a secure water future. https://iwa-network.org/water-reuse-a-critical-step-on-the-path-to-a-secure-water-future/. Assessed 25 Oct 2019
Kaercher JD, Po M, Nancarrow BE (2003) Water recycling community discussion meeting I. Perth: Australian Research Centre for Water in Society (ARCWIS)
Khan SJ, Gerrard LE (2006) Stakeholder communications for successful water reuse operations. Desalination 187(1–3):191–202
Klockner CA (2013) A comprehensive model of the psychology of environmental behaviour—a meta-analysis. Glob Environ Chang 23(5):1028–1038
Leovy J (1997) Reclaimed waste water may ease State's Thirst. Los Angeles Times.
Marlow DR, Moglia M, Cook S, Beale DJ (2013) Towards sustainable urban water management: a critical reassessment. Water Res 47(20):7150–7161
Nancarrow BE, Leviston Z, Po M, Porter NB, Tucker DI (2008) What drives communities' decisions and behaviours in the reuse of wastewater. Water Sci Technol 57(4):485–491
Nancarrow BE, Leviston Z, Tucker DI (2009) Measuring the predictors of communities' behavioural decisions for potable reuse of wastewater. Water Sci Technol 60(12):3199–3209
Ogden J (2003) Some problems with social cognition models: a pragmatic and conceptual analysis. Health Psychol 22(4):424
Poortinga W, Steg L, Vlek C (2004) Values, environmental concern, and environmental behavior: a study into household energy use. Environ Behav 36:70–93
Rivett U, Champanis M, Wilson-Jones T (2013) Monitoring drinking water quality in South Africa: designing information systems for local needs. Water Sa 39(3):409–414
Ross VL, Fielding KS, Louis WR (2014) Social trust, risk perceptions and public acceptance of recycled water: testing a social-psychological model. J Environ Manag 137:61–68
Schwartz SH, Howard JA (1981) A normative decision-making model of altruism. In: Rushton JP, Sorrentino RM (eds) Altruism and helping behavior. Lawerence Erlbaum, Hillsdale, NJ, pp 89–211

Siegrist M, Cvetkovich G, Roth C (2000) Salient value similarity, social trust, and risk/benefit perception. Risk Anal 20:353–362

Stern PC (2000) New environmental theories: toward a coherent theory of environmentally significant behavior. J Soc Issues 56(3):407–424

Taylor DJ, Lichstein KL, Weinstock J, Sanford S, Temple JR (2007) A pilot study of cognitive-behavioral therapy of insomnia in people with mild depression. Behav Ther 38(1):49–57

Thogersen J (1996) Recycling and morality—a critical review of the literature. Environ Behav 28:546–558

UNDP H (2006) Beyond scarcity: power, poverty and the global water crisis. Published for the united nations development program

Webb T, Joseph J, Yardley L, Michie S (2010) Using the internet to promote health behavior change: a systematic review and meta-analysis of the impact of theoretical basis, use of behavior change techniques, and mode of delivery on efficacy. J Med Int Res 12(1):e1376

Wester J, Timpano KR, Cek D, Lieberman D, Fieldstone SC, Broad K (2015) Psychological and social factors associated with wastewater reuse emotional discomfort. J Environ Psychol 42:16–23

Woolston M, Jaffer S (2005) Pricing of water. Occasional paper no.1. Water Services Association of Australia, Melbourne

WWAP (United Nations World Water Assessment Programme) (2015) The United Nations world water development report 2015: water for a sustainable world UNESCO, Paris

Wastewater-Based Epidemiology for Early Warning and Surveillance of Covid-19

Isaac Dennis Amoah, Nashia Deepnarain, Leanne Pillay, Oluyemi Olatunji Awolusi, Sheena Kumari, Taher Abunama, Kriveshin Pillay, and Faizal Bux

1 Introduction

The COVID-19 pandemic, caused by the SARS-CoV-2 virus, has affected several countries and disrupted socio-economic activities globally. As of 20 September 2021, over 228,000,000 people infected and claiming over 4.6 million lives (WHO 2020a). This is the third outbreak caused by coronaviruses in the last 17 years, the first was the severe acute respiratory syndrome coronavirus (SARS-CoV). This emerged in 2003, affecting over 26 countries (Hemida 2019; Lau and Chan 2015; Peiris et al. 2003). The second coronavirus outbreak was the Middle East Respiratory Syndrome Coronavirus (MERS-CoV) that occurred in 2012 (Hemida 2019; Zaki et al. 2012), affecting 27 countries with over 2400 cases as at the end of 2019 (WHO 2020b). The rapid spread of COVID-19 has highlighted several challenges in healthcare services. The major challenge is the effective surveillance of diseases. This has led to a strong focus on developing effective surveillance systems for disease outbreak monitoring. During the COVID-19 pandemic, the focus has been on clinical testing as the main surveillance approach. This approach has been shown to be costly and time-consuming. Wastewater-based epidemiology (WBE) is therefore suggested as a relatively new and cost-effective approach to complement the current, expensive clinical-based surveillance and screening. The WBE tool has immense potential as an early warning system for preventing infectious disease pandemics, especially emerging infectious diseases (Mao et al. 2020). This chapter focuses on the principle of WBE, a discussion of the methodologies applied, and the challenges associated with this tool.

I. D. Amoah · N. Deepnarain · L. Pillay · O. O. Awolusi · S. Kumari · T. Abunama · K. Pillay · F. Bux (✉)
Institute for Water and Wastewater Technology, Durban University of Technology, Durban, South Africa
e-mail: faizalb@dut.ac.za

Y. Li et al. (eds.), *BRICS Countries: Sustainable Water Resource Management and Pollution Control*, https://doi.org/10.1007/978-981-99-9581-3_10

2 Understanding Wastewater-Based Epidemiology (WBE): Principle, History and Current Applications

2.1 *The Principle of Wastewater-Based Epidemiology*

The WBE entails the use of wastewater analysis for screening the presence of biological agents or chemicals for community health monitoring. Substances such as illicit drugs, consumed by individuals, are excreted, either unchanged or as metabolites mixture, whilst infectious agents or their nucleic materials are shed with faeces/urine during infection, which ends up in wastewater sewer networks. The WBE principle postulates that the identified and quantified human biomarkers in untreated wastewater can give an indication of diet, health, diseases and exposure to contaminants within the population connected to the wastewater treatment plant (WWTP) in real time (Gracia-Lor et al. 2017; Hernández et al. 2018).

Chemical or biological agent concentration in untreated wastewater is normalized to per capita mass loads estimated from a combination of daily flow and the total population within the WWTP catchment boundaries (Choi et al. 2018). This offers the possibility of estimating per-capita shedding of the target biological agent or chemicals, its consumption, use and exposure within the WWTP's catchment area. The spatiotemporal assessments of patterns/trends, and response(s) to events within the catchments can be achieved with this approach. This technique includes the estimation at a population scale or within large communities without infringing on individuals' privacy. Since population normalization of data is incorporated, patterns and trends may compare across catchments of different population sizes to determine hotspots (Choi et al. 2018).

2.2 *History of WBE*

The concept of WBE for monitoring illicit drug consumption at the community level whilst preserving an individual's anonymity was first speculated in 2001 (Daughton 2001). Based on systematic sampling and analysis of wastewater samples, it was possible to correlate the drug residues in wastewater to consumption in the community. The first practical implementation of WBE of cocaine consumption monitoring was achieved in 2005 (Mao et al. 2020; Zuccato et al. 2005; Tang et al. 2020a). Since then, WBE has undergone further improvements and has been used for various studies including the presence of pharmaceutical or industrial waste (Baz-Lomba et al. 2016; He et al. 2020), viruses (Hou et al. 2020) and the potential emergence of antibiotic-resistant bacteria (Hutinel et al. 2019; Castrignanò et al. 2020). Substances including alcohol, amphetamines, counterfeit medicine, opiates and tobacco have also been successfully monitored at the community level via WBE (Hummel et al. 2006; Reid et al. 2011; Irvine et al. 2011; Lai et al. 2013; Venhuis et al. 2014; Castiglioni et al. 2015). More recently in 2017, WBE was applied for the first time to estimate human exposure to pesticides across several European cities by quantifying urinary biomarkers in their sewers (Rousis et al. 2017). Different authors have used it to complement other traditional disease surveillance methods successfully in monitoring Norovirus outbreaks in different countries including Australia, New Zealand and Brazil (Lun et al. 2018; Fioretti et al. 2018). In the Netherlands, WBE was applied to establish a timeline for the prevalence of Aichi virus infection in humans (Lodder et al. 2013).

Currently, several countries have reported SARS-CoV-2 detection in wastewater, using it as a data source to determine the circulation of the virus in the community (Ahmed et al. 2020a; Lodder and de Roda Husman 2020; Medema et al. 2020; Wu et al. 2020; Kumar et al. 2020; Randazzo et al. 2020; Rimoldi et al. 2020; La Rosa et al. 2020; Wurtzer et al. 2020). The WBE may provide a better resolution of the infection rate at the community level since all classes of COVID-19 patients (symptomatic, asymptomatic and pre-symptomatic), may not be captured via clinical surveillance. For instance, in the pre-analytical stage, collecting a proper respiratory tract specimen and at the right time from the right anatomical site (thus either the upper or lower respiratory tracts) is considered a major challenge in clinical testing (Tang et al. 2020c). This has led to several challenges with surveillance associated with the resolution of clinical testing.

2.3 Stability of the Virus and Viral RNA in Wastewater

The SARS-CoV-2 virus belongs to the enveloped Coronaviridae family with a positive-sense, single-stranded RNA genome (Naddeo and Liu 2020). They have a distinctive large (20 nm long) club-shaped spike protein (S), the membrane (M) protein, nucleocapsid (N) and the lipid envelope (E). The S, M, and E proteins together form the envelope of the virus (Mousavizadeh and Ghasemi 2020). The association of the nucleocapsid (N) protein with the genomic RNA forms the helical nucleocapsid inside the envelope (Pal et al. 2020). The *structure* of the human coronavirus is shown in Fig. 1.

According to the World Health Organization (WHO 2020c), the SARS-CoV-2 virus is less stable in the environment compared to non-enveloped human enteric viruses with known waterborne transmission (such as adenoviruses, norovirus, rotavirus and hepatitis A virus), due to its enveloped nature. Several studies have reported the inability of the SARS-CoV-2 nucleic materials to persist through the wastewater treatment process. However, lower concentrations of SARS-CoV-2 RNA fragments were also reported from wastewater that was subjected to partial treatment (Wurtzer et al. 2020; Randazzo et al. 2020). The prevailing conditions during wastewater treatment should lead to pathogen reduction through a combination of biological and chemical processes. This was confirmed by other studies that did not detect SARS-CoV-2 in wastewater effluents after tertiary treatment (Randazzo et al. 2020; Sherchan et al. 2020).

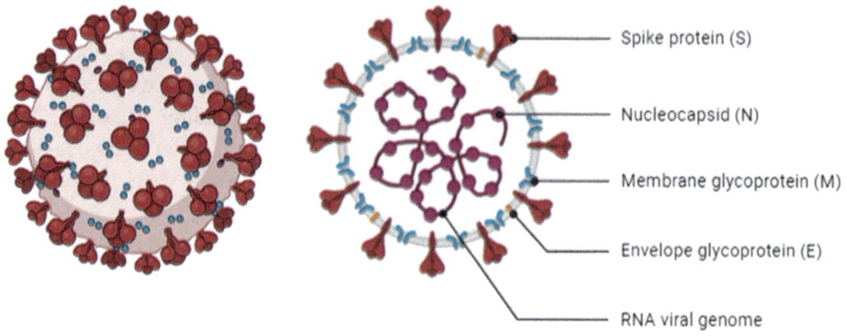

Fig. 1 The human coronavirus structure (Created with BioRender.com)

Despite the widely reported detection of SARS-CoV-2 RNA in wastewater, there are no reports yet on the detection of infective viral particles in wastewater. However, wastewater as a potential route for the transmission of SARS-CoV-2 cannot be ruled out and warrants further investigation. The health risks may be highest in areas of high population density, lack of sufficient wastewater collection and direct exposure of wastewater aerosols (Bogler et al. 2020). Wastewater is a complex matrix consisting of solids, dissolved particulate matter, heavy metals, microorganisms and other macro and micro pollutants. As such, the composition of wastewater could play an important role in the degradation of viruses in the environment. The stability of the viral genome in wastewater should be considered for future environmental surveillance to determine the potential health risks associated with wastewater exposure. The common factors that can affect viral stability include:

- The presence of aerobic microorganisms—Viral RNA is degraded by the action of nucleases secreted by microorganisms (Gundy et al. 2009).
- Viral structure: Enveloped viruses, like coronavirus, have been found to have shorter survival periods compared to nonenveloped viruses and are consistent with the survival of tailed phages (Gundy et al. 2009).
- The presence of organic matter and suspended solids—Viruses tend to attach to solid particles found in wastewater which provides protection as viruses adsorb to these particles. This has been shown to decrease the rate of degradation and inactivation by providing a shield against predation (Ye et al. 2016). The attachment of viral particles to suspended solids in wastewater has been shown to be impacted by pH. An increase in viral adsorption is observed with decreasing pH (Schaub et al. 2017). In addition, pH may also affect the viral stability due to reversible acid denaturation of RNA (Mariani et al. 2018).
- Temperature—Greatly affects the stability and persistence of the virus as higher temperatures denature proteins and increase the activity of extracellular enzymes. Wang et al. (2005) revealed that coronaviruses persist longer in temperatures of 4 °C when compared to 20 °C in hospital wastewater, domestic sewage and

dechlorinated tap water. At 20 °C, coronaviruses persisted in fecal samples for 3 days and urine samples for 17 days.
- Solvents and detergents—Oxidants, such as chlorine, are a common component of household detergents and are readily found in untreated wastewater. Chlorine >0.5 mg/L is sufficient for the complete removal of SARS-CoV in wastewater (Wang et al. 2005). Free chlorine was found to be more effective at inactivating SARS-CoV than chlorine dioxide.

3 Methods Used in the Detection of Coronavirus in Wastewater

The application of WBE for monitoring COVID-19 infections relies on the successful detection of SARS-CoV-2 viral RNA in untreated wastewater. This section therefore describes the current state of knowledge on the processes involved in its detection and quantification from wastewater.

3.1 Sampling, Viral Concentration and RNA Extraction Methods

Several studies postulated that the SARS-CoV-2 virus can be detected in environmental wastewater samples even before the clinical surveillance testing (WHO 2020c; Reilly et al. 2020). Therefore, monitoring of wastewater samples for the early detection and warning may provide critical information, especially in terms of outbreaks in certain areas for the ease of public health and social measures. Suitable methods of sampling, viral concentration and RNA extractions are imperative for the planning of environmental surveillance.

Routine sampling of selective areas such as the WWTP influent, upstream pumping station, biological tanks, final wastewater effluent and/or collection at rivers or latrines are necessary for a widespread environmental surveillance and for early warning (WHO 2020c; Reilly et al. 2020). To reduce areas requiring active public health investigations, more reliable sampling methodologies (especially for areas with low reticulated sewerage coverage), require further investigations (WHO 2020c). The effects of industrial wastewater, runoff from excess rain, and pH of the samples and presence of high concentration of surfactants in domestic wastewater may affect the quality of the samples, which could influence the reliability of the data generated. Accurate estimation of infected populations relies on the use of effective sampling strategies.

Sampling can range from simple batch methodologies (i.e. collection of individual samples), to time integrated sampling and continuous monitoring which closely represent "real-time" monitoring. Grab samples and automated sampling techniques such as conventional submersible in-situ high-frequency auto samplers and conventional refrigerated auto-samplers are currently in use for wastewater sample collection. Most WWTPs are already undergoing automation, primarily for remote control. Further development extends to a wide range of sensors for characterization of wastewater parameters. This could establish an important platform to automate the monitoring in WWTPs and remote reporting by the integration of WBE sensors (Daughton et al. 2020a). The wastewater samples are subsequently transported on ice to the laboratory for analysis. Further processing of wastewater samples in laboratories should follow existing protocols and biosafety standards for proper handling of SARS-CoV-2, i.e. BSL-2 (WHO 2020d). Sampling personnel are required to use standard personal protective equipment (PPE) such as respirators, gowns, safety glasses, gloves, steel-capped boots and hard hats, to protect mucous membranes, skin, airways and clothing from infectious contaminants.

To increase the sensitivity of viral RNA detection, most studies have reported on viral concentration steps. Reliable concentration methods are fundamental for the detection of SARS-Cov-2 in wastewater, which is currently a key research need for WBE (Kitajima et al. 2020; Ahmed et al. 2020a, b; Farkas et al. 2020). All the reports of SARS-CoV-2 in wastewater have been based on the use of viral concentration methods (Ahmed et al. 2020a; Medema et al. 2020; Wu et al. 2020; Nemudryi et al. 2020; Wurtzer et al. 2020; Zhang et al. 2020a, b; Bar Or et al. 2020; Kocamemi et al. 2020; Randazzo et al. 2020; La Rosa et al. 2020).

The wastewater samples generally go through centrifugation and/or filtration to eliminate debris (Farkas et al. 2020). The current methods for concentration of the virus to recover SARS-CoV-2 RNA from wastewater include ultrafiltration (Medema et al. 2020; Wurtzer et al. 2020; Ahmed et al. 2020a), ultracentrifugation (Green et al. 2020a, b; Ahmed et al. 2020b), polyethylene glycol (PEG) precipitation (Wu et al. 2020; Balboa et al. 2020) and skimmed-milk flocculation (Guerrero-Latorre et al. 2020). Globally, electropositive and electronegative membranes based on the electrostatic interaction between the viruses and filters have also been used to concentrate enteric viruses in wastewater samples (Ahmed et al. 2020a; Kitajima et al. 2020). For this method, the viruses which are negatively charged, adsorb onto the filters via salt-bridging. The major disadvantage of these methods is the concentration of organic compounds, which may interfere with downstream processing for further detection and quantitative analysis (Farkas et al. 2020). Most of these methods used for SARS-CoV-2 viral concentration were adopted from methods used for non-enveloped enteric viruses, such as adenovirus, norovirus and enterovirus. Therefore, there is limited information on the recovery efficiencies for the enveloped virus, SARS-CoV-2, which may be different from the nonenveloped enteric viruses due to their structural differences (Ahmed et al. 2020a, b; Medema et al. 2020; Wu et al. 2020). The enveloped viruses, due to the presence of a lipid bilayer surrounding the protein capsid, have a different affinity to non-enveloped viruses, which have a higher tendency to adsorb to solid and/or colloidal particles (Ye et al. 2016; Balboa et al. 2020). Therefore, this suggests that SARS-CoV-2 and the particles are hydrophobic. Such differences may lead to large errors in the estimation of SARS-CoV-2 concentration in wastewater. It is essential to further investigate effective virus concentration methods for the detection of SARS-CoV-2 virus in wastewater (Ahmed et al. 2020a, b).

SARS-CoV-2 viral RNA is commonly extracted from the concentrated viral particles using commercial RNA extraction kits viz. RNeasy PowerMicrobiome Kit (Medema et al. 2020; Ahmed et al. 2020a, b), Biomerieux Nuclisens kit (Medema et al. 2020), PowerFecal Pro kit (Wurtzer et al. 2020) and RNeasy PowerWater Kit and RNeasy (Ahmed et al. 2020a). Furthermore, a list of RNA extraction kits, that may be used for the assessment of SARS-CoV-2 in wastewater samples, was published in the Centres for Disease Control and Prevention of the US (CDC 2020). For the successful application of WBE for early warning and COVID-19 surveillance, further improvements to efficient, rapid and cost-effective concentration and RNA extraction methods are pivotal for monitoring of SARS-CoV-2 and its nucleic acids in wastewater.

3.2 Real Time Quantitative Polymerase Chain Reaction (RT-qPCR)

RT-qPCR is a type of PCR, which is considered quantitative in determining the number of copies of a target gene in a sample. A typical RT-qPCR reaction measures the amount of fluorescence in a sample during each cycle by employing the use of either SYBR Green or TaqMan probe-based chemistries (La Rosa and Muscillo 2013). The standard curve sets the threshold for the reaction, and during the reaction, the fluorescence of each sample is measured (Kuang et al. 2018). If the fluorescence exceeds that of the threshold then it is scored as “positive”. The amount of cycles a sample takes to reach this threshold level is referred to as the cycle threshold or “ct” value. An RT-qPCR reaction can be multiplexed to screen for multiple gene targets within a single sample with ease through the use of specific internal probes that bind to different fluorochromes. To date, RT-qPCR is accepted as the gold standard for the detection of SARS-CoV-2 (Tahamtan and Ardebili 2020). This has been applied to wastewater in various studies with the first being carried out by Ahmed et al. (2020a, b) who used RT-qPCR together with sequencing for the detection of SARS-CoV-2 from pump stations and WWTPs in Australia. This was followed by Wang et al. (2020b) who screened for SARS-CoV-2 in untreated and partially treated wastewater while Nemudryi et al. (2020) used RT-qPCR to monitor raw sewage in municipal WWTPs in the USA. Since the publication of these initial uses of RT-qPCR several other studies (and still counting) have used this approach (Wu et al. 2020; Wurtzer et al. 2020; Lodder and de Roda Husman 2020; Randazzo et al. 2020; Green et al. 2020a, b; Or et al. 2020; Balboa et al. 2020; Peccia et al. 2020; Kumar et al. 2020; Rimoldi et al. 2020).

3.3 Droplet Digital Polymerase Chain Reaction (ddPCR)

Droplet digital PCR is a relatively new surfactant-based platform which allows the user to quantify nucleic acids in a sample with greater precision than qPCR and without the need for a standard calibration curve. In a typical ddPCR reaction, the mastermix is mixed with oil and partitioned into approximately 20,000 nano-sized droplets (using an automated or manual droplet generator) in such a way that each droplet contains little to no copies of the template (either DNA or RNA) (Pinheiro et al. 2012; Quan et al. 2018). A normal PCR amplification step is carried out thereafter using fluorescently labeled TaqMan probes and amplification then proceeds within each encapsulated droplet. The droplets are then passed through an automated reader (similar to a cytometer) which measures the number of positive droplets within each sample well. Poisson statistical analysis is then applied to calculate the absolute concentration of template DNA within the sample and the designated software expresses it in copies per microliter (Roberts et al. 2013). Because the sensitivity of the reaction increases as the partitions increase, the ddPCR platform offers a significant advantage over qPCR in terms of precision and is quite useful in settings that require more robust and accurate quantification of target genes. ddPCR can also be used for gene expression analysis and rare mutation detection (Jamuar et al. 2016). While there is a dearth of information regarding the use of ddPCR for the detection of SARS-CoV-2 in wastewater, a clinical study by Suo et al. (2020) confirmed that ddPCR is superior to qPCR as it has a lower limit of detection (LOD) and is more specific than qPCR when used to screen for SARS-CoV-2 in clinical settings. This method has been used successfully for WBE during the current pandemic (Pillay et al. 2021; Gonzalez et al. 2020; Ciesielski et al. 2021).

3.4 Loop Mediated Isothermal Amplification (LAMP)

This novel nucleotide amplification method allows for the isothermal (at a constant temperature) amplification of target genes within a sample without the need for expensive thermal cycling equipment. This method utilizes multiple primer sets (inner and outer primers) which are designed to create loop structures during the amplification process (Sahoo et al. 2016; Panno et al. 2020). According to Jones (2015), each primer consists of a target-specific section of nucleotides as well as tags of contiguous nucleotides which are not part of the target gene sequence but are only there to allow for loop structures to form. Two types of elongation reactions occur through loop regions during a LAMP reaction. Type 1 is template self-elongation from the stem loop structure formed at three prime terminals and Type 2 is the subsequent binding and elongation of the new primers to the loop region (Notomi et al. 2015). Amplification is indirectly determined by turbidity in sample tubes caused by the production of magnesium pyrophosphate. However, through the addition of fluorescent dyes such as SYBR Green and calcein amongst many others, amplification can also be observed visually through color change (Mori et al. 2001). The use of the LAMP method has revolutionized the way molecular tests are carried out as it can amplify a few copies to up to 10^9 copies in less than an hour with great specificity. The LAMP method does not require the need for skilled personnel and is just as sensitive as other PCR methods (Sahoo et al. 2016). Huang et al. (2020) recognized the need for a simple diagnostic method and developed a LAMP assay to detect 3 genes (orf1 ab, S and N) corresponding to the SARS-CoV-2 genome. In brief, the study was able to visualize a positive result colorimetrically within 30 min and had a limit of detection of 80 copies per millilitre. This was also proven by their qPCR results. Similarly, Baek et al. (2020) developed an RT-LAMP assay for the detection of the N gene of SARS-CoV-2. The LOD was found to be 10^2 copies and there was no cross-reactivity with other viruses further highlighting the specificity of LAMP primers. RT-LAMP has been used to detect SARS-CoV-2 in wastewater and was shown to be a powerful alternative method for the detection of this virus in wastewater (Amoah et al. 2021).

3.5 *Next Generation Sequencing (NGS)*

Next Generation Sequencing (NGS) or "deep sequencing" is a relatively new high-throughput method that provides parallel analysis of samples in a shorter period of time when compared to traditional sequencing methods. In brief, the DNA or RNA to be analyzed is first fragmented and linked to adapters to generate a sequencing library which is then used as a template for the synthesis of numerous short and overlapping nucleic acid fragments (Reuter et al. 2015; Qin 2019). A fluorescently labeled probe attached to a nucleotide is incorporated into each of these fragments making it such that each fragment has its own unique identity. Thereafter, data from all of the sequences are aligned and compared with reference genomes. Each fragment is sequenced numerous times to ensure that the overlapping fragments, thereby reducing error (Voelkerding et al. 2009; Hsiao 2019), cover all regions of the nucleic acid. This coverage is known as the sequencing depth and is affected greatly by the GC content of the sample, whether the sample has pseudogenes, and the presence of mutations (Hsiao 2019).

Using the Illumina sequencing platform, two workflows were developed to sequence SARS-CoV-2 (https://www.nature.com/articles/d42473-020-00120-0). The shotgun metagenomics workflow identifies all microbial communities present in a sample without any prior knowledge as to what microbes are present in the sample, using local data or cloud-based tools. Using this workflow for SARS-CoV-2 analysis allows laboratory workers to determine if patients have other bacterial/viral infections, which allows scientists to track and monitor the SARS-CoV-2 genome for evolutionary traits and to monitor patient responses to SARS-CoV-2. The target enrichment workflow on the other hand only sequences specific genes in a sample but is just as accurate in detecting SARS-CoV-2 as the shotgun workflow. After enrichment, a targeted NGS library and a respiratory virus oligo panel are generated which allows for the sequencing of over 40 pathogens and the detection of genomic mutations in the sample. Although the shotgun workflow has a greater sequencing depth and requires more data, more important information is generated.

3.6 Challenges Associated with Detection of SARS-CoV-2 in Wastewater

3.6.1 The Applicability of Current Viral Concentration Methods

Viral loads in untreated wastewater are relatively low, thereby necessitating the need for viral concentration prior to RNA extraction (Hjelmsø et al. 2017). There are many methods that have been used to concentrate viruses from wastewater however, the majority of these methods were developed for non-enveloped enteric viruses (Haramoto et al. 2018). For example, viral concentration using electropositive membranes is based on the fact that the majority of enteric viruses carry a negative charge in environmental settings (pH neutral) which allows the virus to absorb onto electropositive membranes (Haramoto et al. 2018). A study on coronaviruses by Wang et al. (2005) showed that only 1% of the virus was recoverable from wastewater using current concentration methods. Therefore, there is an urgent need for more intensive research on the applicability of current methods for the concentration of SARS-CoV-2 from wastewater considering that the structural and physical properties of enveloped viruses such as SARS-CoV-2 are much different to that of commonly detected non-enveloped enteric viruses from wastewater (Cascella et al. 2020).

3.6.2 Primer Specificity and Selection of Primers for Detection and Quantification

The selection and design of effective primers that are specific and sensitive enough for the diagnosis and tracking of SARS-CoV-2 are of immense importance, however, it comes with a number of challenges. Primers are the pivotal components of molecular-based assays, but it is only effective provided that mutation and recombination do not occur within the viral genome (Li et al. 2020). The SARS-CoV-2 genome is approximately 30 kb and is arranged in the order of a 5′ untranslated region (UTR)-replicase complex (open reading frame [ORF] 1ab)-structural protein [Spike(S)-Envelope(E)-Membrane (M)-Nucleocapsid (N)]-3′ UTR and non-structural ORFs (Khailany et al. 2020). Proposed gene targets include N (nucleocapsid), E (envelope), ORF 1ab (open reading frame (or RdRP) and S (spike protein genes). The E gene is considered to be the most conserved while the N and RdRP genes are recommended for confirmatory purposes (Chan et al. 2020). Detection of a single target only is said to be prone to error therefore the sensitivity of the assay can be increased by screening for multiple targets (Green et al. 2020a). Although primers and probe designs are evaluated via in-silico analysis, experimental confirmation is of utmost importance. The WHO and the CDC have both published primers and probes for the detection of SARS-CoV-2. The concentration of these primers and probes, reaction conditions, melting temperatures and reverse transcription conditions all need to be optimized for maximum accuracy of the result. Using a positive control, it is important that each primer set be evaluated on the platform it is to be used on, and the absolute limit of detection (ALOD) be determined (Dobnik et al. 2016). Corman et al. (2020) reported ALOD values of 3.8, 5.2 and 8.3 copies per reaction for the detection of RdRp, E and N genes respectively while Shirato et al. (2020) reported that only assays for the N gene worked well. To date, screening for N gene seems to be the most widely used as it produces the most consistent results and has the lowest ALOD.

3.7 The Need for Process Controls

The presence of inhibitors in wastewater samples affects all PCR-based detection and quantification methods (Schrader et al. 2012). Considering that inhibitors may be present in samples and at different concentrations, many studies have recommended the use of process controls in order to monitor the target gene from the viral concentration step to the detection. Two types of process controls exist; the first is known as 'whole process control' and is inoculated into wastewater before any concentration step, and the second is known as the 'microbial process control', which is inoculated prior to RNA extraction (Kitajima et al. 2020). Microbial process control can sometimes also function as a PCR control. The choice of control is important, it must be genetically and structurally similar to the target virus and should not be found naturally in the water sample. To date, the murine hepatitis virus, murine norovirus and mengovirus are acceptable as process controls for the detection of SARS-CoV-2 (Ahmed et al. 2020a, b; Kitajima et al. 2020).

4 Early Warning Based on Detection in Wastewater

The application of WBE as an early warning system for COVID-19 is based on the detection of the viral RNA in feces and subsequently wastewater. The SARS-CoV-2 virus has been reported to infect the glandular cells of the gastrointestinal tract, resulting in gastrointestinal infections (GI) (Yeo et al. 2020; Wan et al. 2020). Early reports indicated that the GI symptoms were observed more frequently in intensive care patients compared to individuals not requiring intensive care (Yang et al. 2020b; Wang et al. 2020a; Chen et al. 2020c). The detection of SARS-CoV-2 RNA in the feces of infected persons is therefore not surprising, based on the GI infections reported. Furthermore, it has been reported that patients with GI symptoms shed higher viral RNA load in their feces compared with other patients without GI infections (Parasa et al. 2020; Chen et al. 2020b; Cheung et al. 2020). Several studies have reported SARS-CoV-2 in feces since the start of the current pandemic (see Table 1). Shedding of viral particles in feces has been reported to occur in the initial period of infection, i.e., just after 1–2 days (with GI infections) and before the onset of pulmonary symptoms (Wang et al. 2020c). However, there is no evidence to show how long shedding continues, as some studies suggested that RNA may be found in the feces between 1 and 25 days after the onset of GI symptoms (Amirian 2020; Ling et al. 2020; Xiao et al. 2020; Zhang et al. 2020a).

The shedding of the viral RNA in feces can be detected in wastewater. Therefore, the principle of early detection of COVID-19 infections through WBE is based on the early detection of the viral RNA signature in wastewater. This may allow for the distributed monitoring of community disease burden for the estimated 2.1 million people connected to over 105, 600 WWTPs globally (Hart and Halden 2020). To cater to the populations not connected to centralized wastewater systems, the concept of sanitation-based epidemiology has been proposed. This approach will allow for expansion of the WBE concept to address non-sewered networks such as those dependent on pit latrine and, urine diversion toilets.

The WBE program for the early tracking of the spread of the SARS-CoV-2 virus can assist water utilities and health authorities since the virus intensity can be detected. According to Hill et al. (2020), WBE surveillance programs have been successfully utilized for the last twenty years to evaluate poliovirus environmental surveillance efficiency. However, deploying such programs for the wastewater surveillance of COVID-19 pandemic, requires not only detecting the virus particle or viral RNA in the community wastewater but also comprehensive wastewater-based datasets interpretation, modeling and online tracking of the wastewater (Fig. 2).

Table 1 Reports of SARS-CoV-2 in fecal material

Sample	Location	References
Feces	Hubei, Shandong and Beijing, China	Wang et al. (2020b)
Feces	Jinhua, China	Zhang et al. (2020a), Wang et al. (2020c)
Anal swabs	Wuhan, China	Zhang et al. (2020b)
Feces	Shanghai and Qingdao, China	Cai et al. (2020)
Feces	Zhuhai, China	Xiao et al. (2020)
Feces	Zhoushan, China	Tang et al. (2020b)
Feces	Singapore	Young et al. (2020)
Feces	Shanghai, China	Ling et al. (2020)
Anal swabs	Guangzhou, China	Chen et al. (2020a)
Feces	Guangdong, China	Chan et al. (2020)
Feces	Kallang, Singapore	Kam et al. (2020)
Feces	Heilongjiang, China	Zhang et al. (2020a)
Feces	China	Xie et al. (2020)
Feces	Macau	Lo et al. (2020)

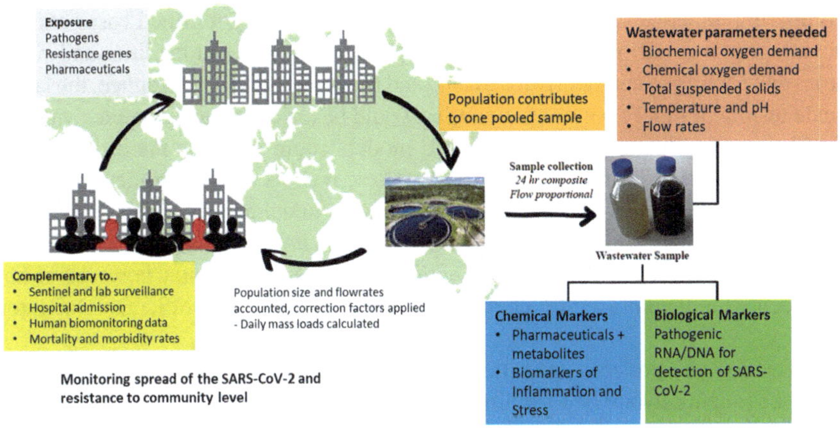

Fig. 2 Representation of wastewater based epidemiology concept (adapted from Sims and Kasprzyk-Hordern (2020))

There are a few studies that may act as a proof of concept for the use of WBE as an early warning system for COVID-19. In a study in the Netherlands, no viral RNA was detected in wastewater samples taken 3 weeks before the first case was reported (Medema et al. 2020). However, almost about a week (6 days) before the first case was reported, SARS-CoV-2 viral RNA was detected. Another study in a Spanish region with the lowest prevalence of COVID-19 detected the viral RNA in wastewater before the first COVID-19 cases were reported (Randazzo et al. 2020). A group of researchers from Yale University in the USA also found the highest concentrations of the SARS-CoV-2 RNA three (3) days before the peak of COVID-19 hospital admissions and seven (7) days before the peak in community cases (Peccia et al. 2020). The most interesting report that may show the potential of WBE as an early warning system is from the state of Santa Catalina in Brazil, where researchers reported that the RNA of SARS-CoV-2 was detected in wastewater samples taken as far back as November 2019 (Fongaro et al. 2020).

These findings provide evidence that WBE for COVID-19 could serve as an early warning system for COVID-19 infections. This is dependent on the use of appropriate methodologies, as discussed above, to ensure that low viral signatures/concentrations are detected.

5 Infection Level Determination in Connected Populations

The utilization of the WBE tool is an effective approach to predict the prevalence of COVID-19 disease in the community by monitoring wastewater (Núñez-delgado 2020). The WBE approach is a necessity, especially in low-income and developing countries, where low confirmed cases of COVID-19 could have been reported comparing to actual cases. This gap between the reported confirmed cases and the actual cases was attributed to a lack of adequate testing kits and facilities and varies from one country to another (Nkengasong and Mankoula 2020). In practice, the large-scale testing of highly populous countries would be expensive and complex, and not feasible to be achieved. Therefore, WBE models could facilitate the estimation of epidemiological indicators from WWTPs. These models could become more accurate by including other related factors such as chemical and biological conditions (Hill et al. 2020). Daughton (2020b) mentioned that WBE could rapidly establish the presence of Covid-19 infections across an entire community; however, it has not been widely embraced by epidemiologists or public health officials (Daughton 2020b).

Detection and possible quantification of SARS-CoV-2 viral RNA in wastewater could be potentially applied for different purposes, such as:

1. direct estimation of disease burden,
2. disease trend analysis,
3. monitoring the efficacy of interventions in reducing diseases,
4. identification of new cases in areas with no known COVID-19 infections.

Different approaches have been proposed in using the viral load in untreated wastewater for the direct estimation of disease burden. For instance, Ahmed et al. (2020a, b) estimated COVID-19 disease burden in Australia using;

$$\text{People in fected} = \frac{\left(\frac{\text{RNA copies}}{\text{liter of wastewater}}\right) * \left(\frac{\text{liters of wastewater}}{\text{day}}\right)}{\left(\frac{\text{gram of feces}}{\text{person per day}}\right) * \left(\frac{\text{RNA copies}}{\text{gram of feces}}\right)}$$

Based on this model, they were able to estimate a median SARS-CoV-2 infection prevalence of 0.096% equivalent to a prevalence of about 450 cases within the communities connected to the WWTP studied.

Vallejo et al. (2020) developed a new model for the estimation of number of cases of COVID-19 with a reliability of almost 90%. This was represented by a linear model of;

$$\text{Number of cases} = -7079 + 1059 * \log V$$

where V is the number of RNA copies per liter of wastewater. They were able to estimate that about 1,661 of the people contributing wastewater to the treatment plant studied were infected. This was very close to the number of clinically confirmed cases reported in that population (1,667 cases).

COVID-19 trends have also been determined by the use of WBE. For instance, Weidhaas et al. (2020), observed that an increase in SARS-CoV-2 RNA in wastewater analyzed corresponded to an outbreak of infections in the connected population. A decline in viral load in untreated wastewater also correlated with a decline in SARS-CoV-2 cases. This approach has been used successfully to determine COVID-19 infection trend in Australia (Ahmed et al. 2020a, b), France (Wurtzer et al. 2020),

USA (Nemudryi et al. 2020; Green et al. 2020b; Peccia et al. 2020), Israel (Or et al. 2020), Turkey (Kocamemi et al. 2020), Spain (Randazzo et al. 2020; Balboa et al. 2020), Netherlands (Medema et al. 2020), India (Kumar et al. 2020), Japan (Hata, et al. 2020), Pakistan (Sharif et al. 2020) and Italy (La Rosa et al. 2020; Rimoldi et al. 2020).

In contrast, Wu et al. (2020) reported a higher viral load in untreated wastewater than expected based on clinical cases. This could be attributed to uncertainties associated with WBE for COVID-19. Goulding et al. (2020) presented three main sources of uncertainty. These are fluctuations in flow, uncertainty in analytical determinations (as discussed above), and the actual size of the population served by the WWTP. Other researchers have used modeling approaches such as the Monte Carlo simulations to address these issues of uncertainty (Croft et al. 2020). These simulations are aimed at reducing the impact of variability in the input parameters required for accurate WBE. Additionally, the viral stability in wastewater may also introduce some levels of uncertainty in WBE. Most publications on WBE for COVID-19 assume complete virus persistence regardless of travel time in the sewer network to the WWTP. This assumption is unrealistic and, therefore could potentially impact the accurate quantification of the viral load. As mentioned above, viral particles in wastewater are affected by several processes that could potentially result in deactivation. To reduce the uncertainty associated with SARS-CoV-2 stability in wastewater the following factors need to be accounted for; in-sewer travel time from point of excretion (shedding) into the sewer network, the temperature and pH within the sewer network (Hart and Halden 2020). Therefore, WBE cannot replace the clinical testing rather it has benefits particularly in informal settlements and low-income areas (Hart and Halden 2020). Table 2 illustrates some applications of WBE for early warning and surveillance of COVID-19 pandemic in BRICS member countries.

6 Conclusion

Wastewater-based epidemiology for the surveillance of COVID-19 infections in connected populations is a cost-effective approach that can be used in resource-constrained regions. This approach is therefore suitable for BRICS member nations and has already seen an increase in interest from member countries such as South Africa, India, Brazil and China. As discussed in this chapter, despite the reports of possible applicability there are a number of challenges that need to be addressed for its effective implementation. These include optimization of sampling approaches, viral concentration and RNA extraction methods and improvement of the models for the prediction of infection numbers. WBE, beyond doubt could play an important role in the surveillance programme of Covid-19 pandemic and in containing and mitigating Covid-19 outbreaks in the communities. However, it is worth noting that in developing countries with no centralized or sewered sanitation systems, expansion of this program to include decentralized sanitation systems will be required. This

Table 2 Application of WBE for early warning and surveillance of COVID-19 pandemic in BRICS member countries

Location/ Country	Description and main conclusions	References
China	The occurrence and potential spread risk of COVID-19 was estimated along the Yangtze, Han, and Fu River basins in Hubei, China. The risk distribution and comparison of watershed cities of COVID-19 were identified. A risk index as proposed, where the distribution order of risk index was Fu River then Han River then Yangtze River. The safety radius and safety time period for the Yangtze, Han, and Fu River are 8 km/14 h, 20 km/30 h and 36 km/36 h. An effective method (modified QMRA index model) was proposed to estimate the potential spread risk of COVID-19 in typical river basins	Yang et al. (2020a)
Sub-Saharan Africa	In this short communication, an early warning wastewater system has been proposed as a platform for SARS-CoV-2 surveillance, and a potentially important public health strategy to combat the disease. It discusses the wastewater surveillance in sub-Saharan Africa and highlights challenges, opportunities and alternatives taken into account the local context	Street et al. (2020)
Brazil	The evidence regarding the presence of the virus in sewage is examined, and a framework was proposed to discuss the disease transmission through water and excreta and how the transmission of a respiratory disease fits into it. The article briefly discusses the practical implications, based on current knowledge, for containment of the pandemic	Heller et al. (2020)
India	The proposed model found significant findings with temporal variation and the viral load in a wastewater treatment plant situated in Ahmedabad, Gujarat, which receives effluent from a hospital treating patients infected with Covid-19. Patterns were obtained from increasing wastewater samples and positive Covid-19 confirmed cases. However, long-term monitoring is necessary for further validation of these findings	Kumar et al. (2020)

will require further optimization of sampling viral concentration and RNA extraction methods to make it applicable to such situations.

References

Ahmed W, Angel N, Edson J, Bibby K, Bivins A, O'Brien JW, Choi PM, Kitajima M, Simpson SL, Li J, Tscharke B (2020a) First confirmed detection of SARS-CoV-2 in untreated wastewater in Australia: A proof of concept for the wastewater surveillance of COVID-19 in the community. Sci Total Environ 728(1):138764

Ahmed W, Bertsch PM, Bivins A, Bibby K, Farkas K, Gathercole A, Haramoto E, Gyawali P, Korajkic A, McMinn BR, Mueller JF, Simpson SL, Smith WJM, Symonds EM, Thomas KV, Verhagen R, Katajima M (2020b) Comparison of virus concentration methods for the RT-qPCR-based recovery of murine hepatitis virus, a surrogate for SARS-CoV-2 from untreated wastewater. Sci Total Environ 739:139960

Amirian ES (2020) Potential fecal transmission of SARS-CoV-2: current evidence and implications for public health. Int J Infect Dis 95:363–370

Amoah ID, Pillay L, Deepnarian N, Awolusi O, Pillay K, Ramlal P, Kumari S, Bux F (2021) Detection of SARSCoV-2 RNA on contact surfaces within shared sanitation facilities. Int J Hyg Environ Health 236:113807. https://doi.org/10.1016/j.ijheh.2021.113807

Baek YH, Um J, Antigua KJC, Park J, Kim Y, Oh S (2020) Development of a reverse transcription-loop-mediated isothermal amplification as a rapid early-detection method for novel SARS-CoV-2. Emerg Microbes Infect 9(1):998–1007

Balboa S, Mauricio-Iglesias M, Rodriguez S, Martínez-Lamas L, Vasallo FJ, Regueiro B, Lema JM (2020) The fate of SARS-COV-2 in WWTPs points out the sludge line as a suitable spot for monitoring. medRxiv. https://doi.org/10.1101/2020.05.25.20112706

Bar Or I, Yaniv K, Shagan M, Ozer E, Erster O, Mendelson E, Mannasse B, Shirazi R, Kramarsky-Winter E, Nir O, Abu-Ali H, Ronen Z, Rinott E, Lewis YE, Friedler EF, Paitan Y, Bitkover E, Berchenko Y, Kushmaro A (2020) Regressing SARS-CoV-2 sewage measurements onto COVID-19 burden in the population: a proof-of-concept for quantitative environmental surveillance

Baz-Lomba JA, Salvatore S, Gracia-Lor E, Bade R, Castiglioni S, Castrignanò E, Causanilles A, Hernandez F, Kasprzyk-Hordern B, Kinyua J, McCall AK (2016) Comparison of pharmaceutical, illicit drug, alcohol, nicotine and caffeine levels in wastewater with sale, seizure and consumption data for 8 European cities. BMC Public Health 16(1):1–11

Bogler A, Packman A, Furman A, Gross A, Kushmaro A, Ronen A, Dagot C, Hill C, Vaizel-Ohayon D, Morgenroth E, Bertuzzo E, Wells G, Kiperwas HR, Horn H, Negev I, Zucker I, Bar-Or I, Moran-Gilad J, Balcazar JL, Bibby K, Elimelech M, Weisbrod N, Nir O, Sued O, Gillor O, Alvarez PJ, Crameri S, Arnon S, Walker S, Yaron S, Nguyen TH, Berchenko Y, Hu Y, Ronen Z, Bar-Zeev E (2020) Rethinking wastewater risks and monitoring in light of the COVID-19 pandemic. Nat Sustain

Cai J, Xu J, Lin D, Xu L, Qu Z, Zhang Y, Zhang H, Jia R, Wang X, Ge Y, Xia A (2020) A case series of children with 2019 novel coronavirus infection: clinical and epidemiological features. Clin Infect Dis 1–5

Cascella M, Rajnik M, Cuomo A, Dulebohn SC, Di Napoli R (2020) Features, evaluation and treatment coronavirus (COVID-19). StatPearls. StatPearls Publishing, Treasure Island (FL)

Castiglioni S, Senta I, Borsotti A, Davoli E, Zuccato E (2015) A novel approach for monitoring tobacco use in local communities by wastewater analysis. Tob Control 24(1):38–42

Castrignanò E, Kannan AM, Proctor K, Petrie B, Hodgen S, Feil EJ, Lewis SE, Lopardo L, Camacho-Muñoz D, Rice J, Cartwright N (2020) (Fluoro) quinolones and quinolone resistance genes in the aquatic environment: a river catchment perspective. Water Res 182(1):116015

CDC-Centers for Disease Control & Prevention (2020) CDC 2019-novel coronavirus (2019-nCoV). real-time RT-PCR diagnostic panel. https://www.fda.gov/media/134922/download. Accessed 25th Aug 2020

Chan JFW, Yip CCY, To KKW, Tang THC, Wong SCY, Leung KH, Fung AYF, Ng ACK, Zou Z, Tsoi HW, Choi GKY, Tam AR, Cheng VCC, Chan KH, Tsang OTY, Yuen KY (2020) Improved molecular diagnosis of COVID-19 by the novel, highly sensitive and specific COVID-19-RdRp/Hel real-time reverse transcription-PCR assay validated in vitro and with clinical specimens. J Clin Microbiol 58(5):1–23

Chen N, Zhou M, Dong X, Qu J, Gong F, Han Y, Qiu Y, Wang J, Liu Y, Wei Y, Yu T (2020a) Epidemiological and clinical characteristics of 99 cases of 2019 novel coronavirus pneumonia in Wuhan, China: a descriptive study. The Lancet 395(10223):507–513

Chen Y, Chen L, Deng Q, Zhang G, Wu K, Ni L, Yang Y, Liu B, Wang W, Wei C, Yang J (2020b) The presence of SARS-CoV-2 RNA in the feces of COVID-19 patients. J Med Virol 92(7):833–840

Chen Y, Guo Y, Pan Y, Zhao ZJ (2020c) Structure analysis of the receptor binding of 2019-nCoV. BiochemBiophys Res Commun. https://doi.org/10.1016/j.bbrc.2020.02.071

Cheung KS, Hung IF, Chan PP, Lung KC, Tso E, Liu R, Ng YY, Chu MY, Chung TW, Tam AR, Yip CC (2020) Gastrointestinal manifestations of SARS-CoV-2 infection and virus load in fecal samples from the Hong Kong cohort and systematic review and meta-analysis. Gastroenterology 159(1):81–95

Choi PM, Tscharke BJ, Donner E, O'Brien JW, Grant SC, Kaserzon SL, Mackie R, O'Malley E, Crosbie ND, Thomas KV, Mueller JF (2018) Wastewater-based epidemiology biomarkers: Past, present and future. Trends Anal Chem 105:453–469

Ciesielski M, Blackwood D, Clerkin T, Gonzalez R, Thompson H, Larson A, Noble R (2021) Assessing sensitivity and reproducibility of RT-ddPCR and RT-qPCR for the quantification of SARS-CoV-2 in wastewater. J Virol Methods 297:114230

Corman VM, Landt O, Kaiser M, Molenkamp R, Meijer A, Chu DKW, Bleicker T, Brünink S, Schneider J, Schmidt ML, Mulders DGJC, Haagmans BL, van der Veer B, van den Brink S, Wijsman L, Goderski G, Romette JL, Ellis J, Zambon M, Peiris M, Goossens H, Reusken C, Koopmans MPG, Drosten C (2020) Detection of 2019 novel coronavirus (2019-nCoV) by real-time RT-PCR. Eurosurveillance 25(3):1–45

Croft TL, Huffines RA, Pathak M, Subedi B (2020) Prevalence of illicit and prescribed neuropsychiatric drugs in three communities in Kentucky using wastewater-based epidemiology and Monte Carlo simulation for the estimation of associated uncertainties. J Hazard Mater 384(1):121306

Daughton CC (2001) Pharmaceuticals and personal care products in the environment: overarching issues and overview. In: Daughton CG, Jones-Lepp TL (eds) ACS symposium series. American Chemical Society, Washington DC, pp 2–38

Daughton CG (2020a) Natural experiment concept to accelerate the Re-purposing of existing therapeutics for Covid-19. Global Epidemiology 2:100026. https://doi.org/10.1016/j.gloepi.2020.100026

Daughton CG (2020b) Wastewater surveillance for population-wide Covid-19: the present and future. Sci Total Environ 736:139631. https://doi.org/10.1016/j.scitotenv.2020.139631

Dobnik D, Štebih D, Blejec A, Morisset D, Žel J (2016) Multiplex quantification of four DNA targets in one reaction with Bio-Rad droplet digital PCR system for GMO detection. Sci Rep 6:35451

Farkas K, Hillary LS, Malham SK, Mcdonald JE, Jones DL (2020) Wastewater and public health: the potential of wastewater surveillance for monitoring COVID-19. Curr Opin Environ Sci Health 17(1):14–20

Fioretti JM, Fumian TM, Rocha MS, dos Santos IDAL, Carvalho-Costa FA, de Assis MR, Rodrigues JDS, Leite JPG, Miagostovich MP (2018) Surveillance of noroviruses in Rio De Janeiro, Brazil: occurrence of new GIV genotype in clinical and wastewater samples. Food Environ Virol 10(1):1–6

Fongaro G, Stoco PH, Souza DSM, Grisard EC, Magri ME, Rogovski P, Schorner MA, Barazzetti FH, Christoff AP, de Oliveira LFV, Bazzo ML (2020) SARS-CoV-2 in human sewage in Santa Catalina, Brazil, Nov 2019. medRxiv

Gonzalez R, Curtis K, Bivins A, Bibby K, Weir MH, Yetka K, Thompson H, Keeling D, Mitchell J, Gonzalez D (2020) COVID-19 surveillance in Southeastern Virginia using wastewater-based epidemiology. Water Res 186:116296

Goulding N, Hickman M, Reid M, Amundsen EJ, Baz-Lomba JA, O'Brien JW, Tscharke BJ, de Voogt P, Emke E, Kuijpers W, Hall W (2020) A comparison of trends in wastewater-based data and traditional epidemiological indicators of stimulant consumption in three locations. Addiction 115(3):462–472

Gracia-Lor E, Castiglioni S, Bade R, Been F, Castrignanò E, Covaci A, González-Mariño I, Hapeshi E, Kasprzyk-Hordern B, Kinyua J, Lai FY, Letzel T, Lopardo L, Meyer MR, O'Brien J, Ramin P, Rousis NI, Rydevik A, Ryu Y, Santos MM, Senta I, Thomaidis NS, Veloutsou S, Yang Z, Zuccato E, Bijlsma L (2017) Measuring biomarkers in wastewater as a new source of epidemiological information: current state and future perspectives. Environ Int 99:131–150

Green DA, Zucker J, Westblade LF, Whittier S, Rennert H, Velu P, Craney A, Cushing M, Liu D, Sobieszczyk ME, Boehme AK, Sepulveda JL (2020a) Clinical performance of SARS-CoV-2 molecular tests. J Clin Microbiol 58(8):e00995-e1020

Green H, Wilder M, Middleton FA, Collins M, Fenty A, Gentile K, Kmush B, Zeng T, Larsen DA (2020b) Quantification of SARS-CoV-2 and cross-assembly phage (crAssphage) from wastewater to monitor coronavirus transmission within communities. medRxiv

Guerrero-Latorre L, Ballesteros I, Villacrés IM, Granda MG, Freire-Paspuel B, Ríos-Touma B (2020) SARS-CoV-2 in river water: implications in low sanitation countries. Sci Total Environ 140832

Gundy PM, Gerba CP, Pepper IL (2009) Survival of Coronaviruses in water and wastewater. Food Environ Virol 1(1):10–14

Haramoto E, Kitajima M, Hata A, Torrey JR, Masago Y, Sano D, Katayama H (2018) A review on recent progress in the detection methods and prevalence of human enteric viruses in water. Water Res 135(1):168–186

Hart OE, Halden RU (2020) Computational analysis of SARS-CoV-2/COVID-19 surveillance by wastewater-based epidemiology locally and globally: feasibility, economy, opportunities and challenges. Sci Total Environ 730(1):138875

Hata A, Honda R, Hara-Yamamura H, Meuchi Y (2020) Detection of SARS-CoV-2 in wastewater in Japan by multiple molecular assays-implication for wastewater-based epidemiology (WBE). medRxiv

He K, Borthwick AG, Lin Y, Li Y, Fu J, Wong Y, Liu W (2020) Sale-based estimation of pharmaceutical concentrations and associated environmental risk in the Japanese wastewater system. Environ Int 139(1):105690

Heller L, Mota CR, Greco DB (2020) COVID-19 faecal-oral transmission: are we asking the right questions? Sci Total Environ 729:138919. https://doi.org/10.1016/j.scitotenv.2020.138919

Hemida MG (2019) Middle East respiratory syndrome coronavirus and the one health concept. PeerJ 7:e7556

Hernández F, Castiglioni S, Covaci A, de Voogt P, Emke E, Kasprzyk-Hordern B, Ort C, Reid M, Sancho JV, Thomas KV, van Nuijs ALN, Zuccato E, Bijlsma L (2018) Mass spectrometric strategies for the investigation of biomarkers of illicit drug use in wastewater. Mass Spectrom Rev 37(3):258–280

Hill K, Zamyadi A, Crosbie ND, Deere D, Vanrolleghem PA (2020) SARS-CoV-2 known and unknowns, implications for the water sector and wastewater-based epidemiology to support national responses worldwide: early review of global experiences with the COVID-19 pandemic. Water Qual Res J 1–11. https://doi.org/10.2166/wqrj.2020.100

Hjelmsø MH, Hellmer M, Fernandez-Cassi F, Timoneda N, Lukjancenko O, Seidel M, Elsasser D, Aarestrup FM, Lofstrom C, Bofill-Mas S, Abril JF, Girones R, Schultz AC (2017) Evaluation of methods for the concentration and extraction of viruses from sewage in the context of metagenomic sequencing. PLoS ONE 1(1):1–17

Hou C, Hua Z, Xu P, Xu H, Wang Y, Liao J, Di B (2020) Estimating the prevalence of hepatitis B by wastewater-based epidemiology in 19 cities in China. Sci Total Environ 740(1):139696

Hsiao SJ (2019) Sources of error in molecular diagnostic analyses. Accurate Results Clin Lab 1(1):337–347

Huang WE, Lim B, Hsu CC, Xiong D, Wu W, Yu Y, Jia H, Wang Y, Zeng Y, Ji M, Chang H, Zhang X, Wang H, Cui Z (2020) RT-LAMP for rapid diagnosis of coronavirus SARS-CoV-2. Microb Biotechnol 13(1):950–961

Hummel D, Löffler D, Fink G, Ternes TA (2006) Simultaneous determination of psychoactive drugs and their metabolites in aqueous matrices by liquid chromatography mass spectrometry. Environ Sci Technol 40(23):7321–7328

Hutinel M, Huijbers PMC, Fick J, Åhrén C, Larsson DGJ, Flach CF (2019) Population-level surveillance of antibiotic resistance in Escherichia coli through sewage analysis. Eurosurveillance 24(37):1800497

Irvine RJ, Kostakis C, Felgate PD, Jaehne EJ, Chen C, White JM (2011) Population drug use in Australia: a wastewater analysis. Forensic Sci Int 210:69–73

Jamuar SS, D'Gama AM, Walsh CA (2016) Somatic mosaicism and neurological diseases. In: Lehner T, Miller BL, State MW (eds) Genomics, circuits, and pathways in clinical neuropsychiatry. Academic Press, USA, pp 179–199

Jones JL (2015) High throughput screening of seafood for foodborne pathogens. In: Bhunia AK, Kim MS, Taitt CR (ed) Biosensor technologies, hyperspectral imaging and practical applications. Woodhead Publishing, UK

Kam KQ, Yung CF, Cui L, Lin RTP, Mak TM, Maiwald M, Li J, Chong CY, Nadua K, Tan NWH, Thoon KC (2020) A well infant with Coronavirus disease 2019 with high viral load. Clin Infect Dis 71(15):847–849

Khailany RA, Safdar M, Ozaslan M (2020) Genomic characterization of a novel SARS-CoV-2. Gene Rep 19(1):100682

Kitajima M, Ahmed W, Bibby K, Carducci A, Gerba CP, Hamilton KA, Haramoto E, Rose JB (2020) SARS-CoV-2 in wastewater: state of the knowledge and research needs. Sci Total Environ 739(1):139076

Kocamemi BA, Kurt H, Sait A, Sarac F, Saatci AM, Pakdemirli B (2020) SARS-CoV-2 detection in Istanbul wastewater treatment plant sludges. medRxiv

Kuang J, Yan X, Genders AJ, Granata C, Bishop DJ (2018) An overview of technical considerations when using quantitative real-time PCR analysis of gene expression in human exercise research. PLoS ONE 13(5):e0196438

Kumar M, Patel AK, Shah AV, Raval J, Rajpara N, Joshi M, Joshi CG (2020) First proof of the capability of wastewater surveillance for COVID-19 in India through detection of genetic material of SARS-CoV-2. Sci Total Environ 746(1):141326

La Rosa G, Iaconelli M, Mancini P, Ferraro GB, Veneri C, Bonadonna L, Lucentini L, Suffredini E (2020) First detection of SARS-CoV-2 in untreated wastewaters in Italy. Sci Total Environ 736(1):139652

Lai FY, Bruno R, Leung HW, Thai PK, Ort C, Carter S, Thompson K, Lam PK, Mueller JF (2013) Estimating daily and diurnal variations of illicit drug use in Hong Kong: a pilot study of using wastewater analysis in an Asian metropolitan city. Forensic Sci Int 233(1–3):126–132

Lau SK, Chan JF (2015) Coronaviruses: emerging and re-emerging pathogens in humans and animals. Virol J 12:209. https://doi.org/10.1186/s12985-015-0432-z

Li D, Zhang J, Li J (2020) Primer design for quantitative real-time PCR for the emerging Coronavirus SARS-CoV-2. Theranostics 10(16):7150–7162

Ling Y, Xu S, Lin Y, Tian D, Zhu Z, Dai F, Wu F, Song Z, Huang W, Chen J, Hu B (2020) The persistence and clearance of viral RNA in 2019 novel coronavirus disease survivors. Chin Med J 133(9):1039–1043

Lo IL, Lio CF, Cheong HH, Lei CI, Cheong TH, Zhong X, Tian Y, Sin NN (2020) Evaluation of SARS-CoV-2 RNA shedding in clinical specimens and clinical characteristics of 10 patients with COVID-19 in Macau. Int J Biol Sci 16(10):1698

Lodder W, de Roda Husman AM (2020) SARS-CoV-2 in wastewater: potential health risk, but also data source. The lancet. Gastroenterol Hepatol 1253(20):30087

Lodder WJ, Rutjes SA, Takumi K, de Roda Husman AM (2013) Aichi virus in sewage and surface water, the Netherlands. Emerg Infect Dis 19(8):1222–1230

Lun JH, Hewitt J, Sitabkhan A, Eden JS, Enosi Tuipulotu D, Netzler NE, Morrell L, Merif J, Jones R, Huang B, Warrilow D, Ressler KA, Ferson MJ, Dwyer DE, Kok J, Rawlinson WD, Deere D, Crosbie ND, White PA (2018) Emerging recombinant noroviruses identified by clinical and wastewater screening article. Emerg Microbes Infect 7(1):1–14

Mao K, Zhang K, Du W, Ali W, Feng X, Zhang H (2020) The potential of wastewater-based epidemiology as surveillance and early warning of infectious disease outbreaks. Curr Opin Environ Sci Health 17:1–7

Mariani A, Bonfio C, Johnson CM, Sutherland JD (2018) PH-Driven RNA strand separation under prebiotically plausible conditions. Biochemistry 57(45):6382–6386

Medema G, Heijnen L, Elsinga G, Italiaander R, Brouwer A (2020) Presence of SARS-Coronavirus-2 RNA in sewage and correlation with reported COVID-19 prevalence in the early stage of the epidemic in the Netherlands. Environ Sci Technol Lett 7(7):511–516

Mori Y, Nagamine K, Tomita N, Notomi T (2021) Detection of loop-mediated isothermal amplification reaction by turbidity derived from magnesium pyrophosphate formation. Biochem Biophys Res Commun 289(1):150–154

Mousavizadeh L, Ghasemi S (2020) Genotype and phenotype of COVID-19: their roles in pathogenesis. J Microbiol Immunol Infect. https://doi.org/10.1016/j.jmii.2020.03.022

Naddeo V, Liu H (2020) Editorial Perspectives: 2019 novel coronavirus (SARS-CoV-2): what is its fate in urban water cycle and how can the water research community respond? Environ Sci: Water Res Technol 6(5):1213–1216

Nemudryi A, Nemudraia A, Surya K, Wiegand T, Buyukyoruk M, Wilkinson R, Wiedenheft B (2020) Temporal detection and phylogenetic assessment of SARS-CoV-2 in municipal wastewater. medRxiv

Nkengasong JN, Mankoula W (2020) Comment Looming threat of COVID-19 infection in Africa: act collectively, and fast. Lancet 395(10227):841–842. https://doi.org/10.1016/S0140-6736(20)30464-5

Notomi T, Mori Y, Tomita N, Kanda H (2015) Loop-mediated isothermal amplification (LAMP): principle, features, and future prospects. J Microbiol 52(1):1–5

Núñez-delgado A (2020) Science of the Total Environment What do we know about the SARS-CoV-2 coronavirus in the environment ? Sci Total Environ 727:138647. https://doi.org/10.1016/j.scitotenv.2020.138647

Or IB, Yaniv K, Shagan M, Ozer E, Erster O, Mendelson E, Mannasse B, Shirazi R, Kramarsky-Winter E, Nir O, Abu-Ali H (2020) Regressing SARS-CoV-2 sewage measurements onto COVID-19 burden in the population: a proof-of-concept for quantitative environmental surveillance. medRxiv

Pal M, Berhanu G, Desalegn C, Kandi V (2020) Severe acute respiratory syndrome Coronavirus-2 (SARS-CoV-2): an update. Cureus 12(3):7423. https://doi.org/10.7759/cureus.7423

Panno S, Matic S, Tiberini A, Caruso AG, Bella P, Torta L, Stassi R, Davino S (2020) Loop mediated isothermal amplification: principles and applications in plant virology. Plants 9(4):461

Parasa S, Desai M, Chandrasekar VT, Patel HK, Kennedy KF, Roesch T, Spadaccini M, Colombo M, Gabbiadini R, Artifon EL, Repici A (2020) Prevalence of gastrointestinal symptoms and fecal viral shedding in patients with coronavirus disease 2019: a systematic review and meta-analysis. JAMA Netw Open 3(6):e2011335–e2011335

Peccia J, Zulli A, Brackney DE, Grubaugh ND, Kaplan EH, Casanovas-Massana A, Ko AI, Malik AA, Wang D, Wang M, Weinberger DM (2020) SARS-CoV-2 RNA concentrations in primary municipal sewage sludge as a leading indicator of COVID-19 outbreak dynamics. medRxiv

Peiris JSM, Lai ST, Poon LLM, Guan Y, Yam LYC, Lim W, Nicholls J, Yee WKS, Yan WW, Cheung MT, Cheng VCC (2003) Coronavirus as a possible cause of severe acute respiratory syndrome. Lancet 361(9366):1319–1325

Pillay L, Amoah ID, Deepnarain N, Pillay K, Awolusi OO, Kumari S, Bux F (2021) Monitoring changes in COVID-19 infection using wastewater-based epidemiology: A South African perspective. Sci Total Environ 786:147273. https://doi.org/10.1016/j.scitotenv.2021.147273

Pinheiro LB, Coleman VA, Hindson CM, Herrmann J, Hindson BJ, Bhat S, Emslie KR (2012) Evaluation of a droplet digital polymerase chain reaction format for DNA copy number quantification. Anal Chem 84(1):1003–1011

Qin D (2019) Next-generation sequencing and its clinical application. Cancer Biol Med 16(1):4–10

Quan PL, Sauzade M, Brouzes E (2018) DPCR: a technology review. Sensors 18(1):1–27

Randazzo W, Truchado P, Cuevas-Ferrando E, Simón P, Allende A, Sánchez G (2020) SARS-CoV-2 RNA in wastewater anticipated COVID-19 occurrence in a low prevalence area. Water Res 118(1):115942

Reid MJ, Langford KH, Mørland J, Thomas KV (2011) Analysis and interpretation of specific ethanol metabolites, ethyl sulfate, and ethyl glucuronide in sewage effluent for the quantitative measurement of regional alcohol consumption. Alcohol: Clin Exp Res 35(9):1593–1599

Reilly KMO, Allen DJ, Fine P, Asghar H (2020) The challenges of informative wastewater sampling for SARS-CoV-2 must be met: lessons from polio eradication. Lancet Microbe. https://doi.org/10.1016/S2666-5247(20)30100-2

Reuter JA, Spacek DV, Snyder MP (2015) High-throughput sequencing technologies. Mol Cell 58(4):586–597

Rimoldi SG, Stefani F, Gigantiello A, Polesello S, Comandatore F, Mileto D, Maresca M, Longobardi C, Mancon A, Romeri F, Pagani C, Cappelli F, Roscioli C, Moja L, Gismondo MR, Salerno F (2020) Presence and infectivity of SARS-CoV-2 virus in wastewaters and rivers. Sci Total Environ 744(1):140911

Roberts CH, Last A, Molina-Gonzale S, Cassama E, Butcher R, Nabicassa M, McCarthy E, Burr SE, Mabey DC, Bailey RL, Holland MJ (2013) Development and evaluation of a next-generation digital PCR diagnostic assay for ocular Chlamydia trachomatis infections. J Clin Microbiol 51(7):2195–2203

La Rosa G, Muscillo M (2013) Molecular detection of viruses in water and sewage. In: Cook N (ed) Viruses in food and water: risks, surveillance and control. Woodhead Publishing, India

Rousis NI, Gracia-Lor E, Zuccato E, Bade R, Baz-Lomba JA, Castrignanò E, Causanilles A, Covaci A, de Voogt P, Hernàndez F, Kasprzyk-Hordern B, Kinyua J, McCall AK, Plósz BG, Ramin P, Ryu Y, Thomas KV, van Nuijs A, Yang Z, Castiglioni S (2017) Wastewater-based epidemiology to assess pan-European pesticide exposure. Water Res 121:270–279

Sahoo PR, Sethy K, Mohapatra S, Panda D (2016) Loop mediated isothermal amplification: an innovative gene amplification technique for animal diseases. Vet World 9(5):465–469

Schaub SA, Sorber CA, Taylor GW, Lefler E, Kott Y, Malina JF, Ranganathan KR, Moore BED, Sagik BP, Nupen EM, Bateman BW, McKenny NC, Gerba CP, Sobsey MD, Wallis C (2017) Virus survival in water and wastewater systems. Tex Univ, Cent for Res in Water Resour, Symp Ser, no 7

Schrader C, Schielke A, Ellerbroek L, Johne R (2012) PCR inhibitors—occurrence, properties and removal. J Appl Microbiol 113(5):1014–1026

Sharif S, Ikram A, Khurshid A, Salman M, Mehmood N, Arshad Y, Ahmad J, Angez M, Alam MM, Rehman L, Mujtaba G (2020) Detection of SARS-Coronavirus-2 in wastewater, using the existing environmental surveillance network: an epidemiological gateway to an early warning for COVID-19 in communities. medRxiv

Sherchan SP, Shahin S, Ward LM, Tandukar S, Aw TG, Schmitz B, Ahmed W, Kitajima M (2020) First detection of SARS-CoV-2 RNA in wastewater in North America: a study in Louisiana, USA. Sci Total Environ 743:140621

Sims N, Kasprzyk-Hordern B (2020) Future perspectives of wastewater-based epidemiology: monitoring infectious disease spread and resistance to the community level. Environ Int 139:105689

Shirato K, Nao N, Katano H, Takaymam I, Saito S, Kato F, Katoh H, Sakata M, Nakatsu Y, Mori Y, Kageyama T, Matsuyama S, Takeda M (2020) Development of genetic diagnostic methods for detection for novel coronavirus 2019(nCoV-2019) in Japan. Jpn J Infect Dis 73(4)

Street R, Malema S, Mahlangeni N, Mathee A (2020) Science of the total environment wastewater surveillance for Covid-19: an African perspective. Sci Total Environ 743:140719. https://doi.org/10.1016/j.scitotenv.2020.140719

Suo T, Liu X, Feng J, Guo M, Hu M, Guo D, Ullah H, Yang Y, Zhang Q, Wang X, Sajid M, Huang Z, Deng L, Chen T, Liu F, Xu K, Zhang Q, Liu Y, Xiong Y, Chen G, Lan K, Chen Y (2020) DDPCR: a more accurate tool for SARS-CoV-2 detection in low viral load specimens. Emerg Microbes Infect 9(1):159–168

Tahamtan A, Ardebili A (2020) Real-time RT-PCR in COVID-19 detection: issues affecting the results. Expert Rev Mol Diagn 20(5):453–454

Tang S, He C, Thai PK, He A, Vijayasarathy S, Toms L, Thompson K, Hobson P, Tscharke BJ, Brien JWO, Thomas KV, Mueller JF (2020a) Urinary concentrations of bisphenols in the Australian population and their association with the per capita mass loads in wastewater. Environ Sci Technol 1–8

Tang A, Tong ZD, Wang HL, Dai YX, Li KF, Liu JN, Wu WJ, Yuan C, Yu ML, Li P, Yan JB (2020b) Early release-detection of novel Coronavirus by RT-PCR in stool specimen from asymptomatic child, China. Emerg Infect Dis 26(6)

Tang YW, Schmitz JE, Persing DH, Stratton CW (2020c) Laboratory diagnosis of COVID-19: current issues and challenges. J Clin Microbiol 58(6)

Vallejo JA, Rumbo-Feal S, Conde-Pérez K, López-Oriona Á, Tarrío J, Reif R, Ladra S, Rodiño-Janeiro BK, Nasser M, Cid Á, Veiga MC, Acevedo A, Lamora C, Bou G, Cao R, Poza M (2020) Highly predictive regression model of active cases of COVID-19 in a population by screening wastewater viral load. medRxiv. https://doi.org/10.1101/2020.07.02.20144865

Venhuis BJ, Voogt P, Emke E, Causanilles A, Keizers PHJ (2014) Success of rogue online pharmacies: sewage study of sildenafil in the Netherlands. BMJ 349:g4317

Voelkerding KV, Dames SA, Durtschi JD (2009) Next-generation sequencing: from basic research to diagnostics. Clin Chem 55(4):641–658

Wan Y, Shang J, Graham R, Baric RS, Li F (2020) Receptor recognition by the novel coronavirus from Wuhan: an analysis based on decade-long structural studies of SARS coronavirus. J Virol 94(7):1–9

Wang XW, Li J, Guo T, Zhen B, Kong Q, Yi B, Li Z, Song N, Jin M, Xiao W, Zhu X, Gu C, Yin J, Wei W, Yao W, Liu C, Li J, Ou G, Wang M, Fang T, Wang G, Qiu Y, Wu H, Chao F, Li J (2005) Concentration and detection of SARS coronavirus in sewage from Xiao Tang Shan hospital and the 309th Hospital of the Chinese People's Liberation Army. Water Sci Technol 52(8):213–221

Wang D, Hu B, Hu C, Zhu F, Liu X, Zhang J, Wang B, Xiang H, Cheng Z, Xiong Y, Zhao Y, Li Y, Wang X, Peng Z (2020a) Clinical characteristics of 138 hospitalized patients with 2019 novel coronavirus-infected pneumonia in Wuhan, China. JAMA 323(11):1061–1069

Wang J, Feng H, Zhang S, Ni Z, Chen Y, Zhuo L, Zhong Z, Qu T (2020b) SARS-CoV-2 RNA detection of hospital isolation wards hygiene monitoring during the Coronavirus Disease 2019 outbreak in a Chinese hospital. Int J Infect Dis 94(1):103–106

Wang W, Xu Y, Gao R, Lu R, Han K, Wu G, Tan W (2020c) Detection of SARS-CoV-2 in different types of clinical specimens. JAMA 323(18):1843–1844

Weidhaas J, Aanderud Z, Roper D, VanDerslice J, Gaddis E, Ostermiller J, Hoffman K, Jamal R, Heck P, Zhang Y, Torgersen K (2020) Correlation of SARS-CoV-2 RNA in wastewater with COVID-19 disease burden in sewersheds. https://doi.org/10.21203/rs.3.rs-40452/v1

WHO-World Health Organization (2020b) Middle East respiratory syndrome coronavirus (MERS-CoV). https://www.who.int/emergencies/mers-cov/en/. Accessed 4 Sep 2020

WHO-World Health Organization (2020c) Water, sanitation, hygiene, and waste management for SARS-CoV-2, the virus that causes COVID-19: interim guidance, 29 July 2020 (no. WHO/COVID-19/IPC_WASH/2020.4). World Health Organization. https://apps.who.int/iris/bitstream/handle/10665/333560/WHO-2019-nCoV-IPC_WASH-2020.4-eng.pdf?sequence=10&isAllowed=y

WHO-World Health Organization (2020d) Laboratory biosafety guidance related to coronavirus disease (COVID-19). https://www.who.int/publications/i/item/laboratory-biosafety-guidance-related-to-coronavirus-disease-(covid-19). Accessed 29 Jun 2020

WHO-World Health Organization (2020a) Coronavirus disease (COVID-19) Situation report. https://www.who.int/docs/default-source/coronaviruse/situation-reports/20200704-covid-19-sitrep-166.pdf?sfvrsn=6247972_2. Accessed 4 Sep 2020

Wu F, Xiao A, Zhang J, Gu X, Lee WL, Kauffman K, Hanage W, Matus M, Ghaeli N, Endo N, Duvallet C, Moniz K, Erickson T, Chai P, Thompson J, Alm E (2020) SARS-CoV-2 titers in wastewater are higher than expected from clinically confirmed cases.https://doi.org/10.1101/2020.04.05.20051540

Wurtzer S, Marechal V, Mouchel JM, Moulin L (2020) Time course quantitative detection of SARS-CoV-2 in Parisian wastewaters correlates with COVID-19 confirmed cases.https://doi.org/10.1101/2020.04.12.20062679

Xiao F, Tang M, Zheng X, Liu Y, Li X, Shan H (2020) Evidence for gastrointestinal infection of SARS-CoV-2. Gastroenterology 158(6):1831–1833

Xie C, Jiang L, Huang G, Pu H, Gong B, Lin H, Ma S, Chen X, Long B, Si G, Yu H (2020) Comparison of different samples for 2019 novel coronavirus detection by nucleic acid amplification tests. Int J Infect Dis 93(1):264–267

Yang B, Li W, Wang J, Tian Z, Cheng X, Zhang Y et al (2020a) Science of the Total Environment Estimation of the potential spread risk of COVID-19: occurrence assessment along the Yangtze, Han, and Fu River basins in Hubei, China. Sci Total Environ 746:141353. https://doi.org/10.1016/j.scitotenv.2020.141353

Yang J, Zheng Y, Gou X, Pu K, Chen Z, Guo Q, Ji R, Wang H, Wang Y, Zhou Y (2020b) Prevalence of comorbidities in the novel Wuhan coronavirus (COVID-19) infection: a systematic review and meta-analysis. Int J Infect Dis 94(1):91–95

Ye Y, Ellenberg RM, Graham KE, Wigginton KR (2016) Survivability, partitioning, and recovery of enveloped viruses in untreated municipal wastewater. Environ Sci Technol 50:5077–5085

Yeo C, Kaushal S, Yeo D (2020) Enteric involvement of coronaviruses: is faecal–oral transmission of SARS-CoV-2 possible? Lancet Gastroenterol Hepatol 5(4):335–337

Young BE, Ong SWX, Kalimuddin S, Low JG, Tan SY, Loh J, Ng OT, Marimuthu K, Ang LW, Mak TM, Lau SK et al (2020) Epidemiologic features and clinical course of patients infected with SARS-CoV-2 in Singapore. JAMA 323(15):1488–1494

Zaki AM, Van Boheemen S, Bestebroer TM, Osterhaus AD, Fouchier RA (2012) Isolation of a novel coronavirus from a man with pneumonia in Saudi Arabia. N Engl J Med 367(19):1814–1820

Zhang J, Wang S, Xue Y (2020a) Fecal specimen diagnosis 2019 novel coronavirus–infected pneumonia. J Med Virol 92(6):680–682

Zhang Y, Chen C, Zhu S, Shu C, Wang D, Song J, Song Y, Zhen W, Zijian F, Wu G, Xu J (2020b) Isolation of 2019-nCoV from a stool specimen of a laboratory confirmed case of the coronavirus disease 2019 (COVID-19). Environ Int 8:123–124

Zuccato E, Chiabrando C, Castiglioni S, Calamari D, Bagnati R, Schiarea S, Fanelli R (2005) Cocaine in surface waters: a new evidence-based tool to monitor community drug abuse. Environ Health 4(1):14–21

GPSR Compliance

The European Union's (EU) General Product Safety Regulation (GPSR) is a set of rules that requires consumer products to be safe and our obligations to ensure this.

If you have any concerns about our products, you can contact us on ProductSafety@springernature.com

In case Publisher is established outside the EU, the EU authorized representative is:

Springer Nature Customer Service Center GmbH
Europaplatz 3
69115 Heidelberg, Germany

Batch number: 08250931

Printed by Printforce, the Netherlands